Student Study Guide to Accompany
Statistics
Alive!

Student Study Guide to Accompany

Statistics Alive!

Wendy J. Steinberg
Affiliate Faculty, Division of Educational Psychology and Methodology, University at Albany, SUNY

Student Study Guide material created by Matthew Price
Georgia State Psychology Department

SAGE Publications
Los Angeles • London • New Delhi • Singapore

Copyright © 2008 by Sage Publications, Inc.

All rights reserved. No part of this book may be reproduced or utilized in any form or by any means, electronic or mechanical, including photocopying, recording, or by any information storage and retrieval system, without permission in writing from the publisher.

For information:

Sage Publications, Inc.
2455 Teller Road
Thousand Oaks, California 91320
E-mail: order@sagepub.com

Sage Publications Ltd.
1 Oliver's Yard
55 City Road
London EC1Y 1SP
United Kingdom

Sage Publications India Pvt. Ltd.
B 1/I 1 Mohan Cooperative Industrial Area
Mathura Road, New Delhi 110 044
India

Sage Publications Asia-Pacific Pte. Ltd.
33 Pekin Street #02-01
Far East Square
Singapore 048763

Printed in the United States of America

This book is printed on acid-free paper.

08 09 10 11 12 10 9 8 7 6 5 4 3 2

Acquisitions Editor:	Cheri Dellelo/Jerry Westby
Associate Editor:	Deya Saoud
Editorial Assistant:	Lara Grambling
Production Editor:	Laureen A. Shea
Typesetter:	C&M Digitals (P) Ltd.
Cover Designer:	Bryan Fishman
Marketing Manager:	Stephanie Adams

Contents

Part I Study Guide. Preliminary Information — 1
    Part I Summary — 1
    Learning Objectives — 3
    Computational Exercises — 3
    Answers to Odd-Numbered Computational Exercises — 5
    True/False Questions — 6
    Answers to Odd-Numbered True/False Questions — 7
    Short-Answer Questions — 7
    Answers to Odd-Numbered Short-Answer Questions — 8
    Multiple-Choice Questions — 8
    Answers to Odd-Numbered Multiple-Choice Questions — 11

Part II Study Guide. Tables and Graphs — 13
    Part II Summary — 13
    Learning Objectives — 15
    Computational Exercises — 15
    Answers to Odd-Numbered Computational Exercises — 17
    True/False Questions — 18
    Answers to Odd-Numbered True/False Questions — 19
    Short-Answer Questions — 19
    Answers to Odd-Numbered Short-Answer Questions — 20
    Multiple-Choice Questions — 20
    Answers to Odd-Numbered Multiple-Choice Questions — 24

Part III Study Guide. Central Tendency — 25
    Part III Summary — 25
    Learning Objectives — 26
    Computational Exercises — 26
    Answers to Odd-Numbered Computational Exercises — 27
    True/False Questions — 28
    Answers to Odd-Numbered True/False Questions — 28
    Short-Answer Questions — 28
    Answers to Odd-Numbered Short-Answer Questions — 29
    Multiple-Choice Questions — 30
    Answers to Odd-Numbered Multiple-Choice Questions — 33

Part IV Study Guide. Dispersion — 35
    Part IV Summary — 35
    Learning Objectives — 36
    Computational Exercises — 36
    Answers to Odd-Numbered Computational Exercises — 38

True/False Questions	38
Answers to Odd-Numbered True/False Questions	39
Short-Answer Questions	39
Answers to Odd-Numbered Short-Answer Questions	40
Multiple-Choice Questions	41
Answers to Odd-Numbered Multiple-Choice Questions	44

Part V Study Guide. The Normal Curve and Standard Scores — 45

Part V Summary	45
Learning Objectives	47
Computational Exercises	47
Answers to Odd-Numbered Computational Exercises	49
True/False Questions	49
Answers to Odd-Numbered True/False Questions	50
Short-Answer Questions	50
Answers to Odd-Numbered Short-Answer Questions	51
Multiple-Choice Questions	52
Answers to Odd-Numbered Multiple-Choice Questions	55

Part VI Study Guide. Probability — 57

Part VI Summary	57
Learning Objectives	59
Computational Exercises	59
Answers to Odd-Numbered Computational Exercises	61
True/False Questions	62
Answers to Odd-Numbered True/False Questions	62
Short-Answer Questions	63
Answers to Odd-Numbered Short-Answer Questions	64
Multiple-Choice Questions	64
Answers to Odd-Numbered Multiple-Choice Questions	67

Part VII Study Guide. Inferential Theory — 69

Part VII Summary	69
Learning Objectives	71
Computational Exercises	72
Answers to Odd-Numbered Computational Exercises	73
True/False Questions	74
Answers to Odd-Numbered True/False Questions	74
Short-Answer Questions	75
Answers to Odd-Numbered Short-Answer Questions	75
Multiple-Choice Questions	76
Answers to Odd-Numbered Multiple-Choice Questions	79

Part VIII Study Guide. The One-Sample Test — 81

Part VIII Summary	81
Learning Objectives	84
Computational Exercises	85
Answers to Odd-Numbered Computational Exercises	87
True/False Questions	87
Answers to Odd-Numbered True/False Questions	88
Short-Answer Questions	88
Answers to Odd-Numbered Short-Answer Questions	89
Multiple-Choice Questions	90
Answers to Odd-Numbered Multiple-Choice Questions	93

Part IX Study Guide. The Two-Sample Test   95
    Part IX Summary   95
    Learning Objectives   98
    Computational Exercises   99
    Answers to Odd-Numbered Computational Exercises   103
    True/False Questions   105
    Answers to Odd-Numbered True/False Questions   105
    Short-Answer Questions   106
    Answers to Odd-Numbered Short-Answer Questions   107
    Multiple-Choice Questions   108
    Answers to Odd-Numbered Multiple-Choice Questions   110

Part X Study Guide. The Multisample Test   111
    Part X Summary   111
    Learning Objectives   114
    Computational Exercises   114
    Answers to Odd-Numbered Computational Exercises   119
    True/False Questions   120
    Answers to Odd-Numbered True/False Questions   121
    Short-Answer Questions   121
    Answers to Odd-Numbered Short-Answer Questions   122
    Multiple-Choice Questions   122
    Answers to Odd-Numbered Multiple-Choice Questions   125

Part XI Study Guide. Post Hoc Tests   127
    Part XI Summary   127
    Learning Objectives   128
    Computational Exercises   128
    Answers to Odd-Numbered Computational Exercises   133
    True/False Questions   134
    Answers to Odd-Numbered True/False Questions   135
    Short-Answer Questions   135
    Answers to Odd-Numbered Short-Answer Questions   136
    Multiple-Choice Questions   137
    Answers to Odd-Numbered Multiple-Choice Questions   141

Part XII Study Guide. More Than One Independent Variable   143
    Part XII Summary   143
    Learning Objectives   145
    Computational Exercises   146
    Answers to Odd-Numbered Computational Exercises   152
    True/False Questions   154
    Answers to Odd-Numbered True/False Questions   155
    Short-Answer Questions   155
    Answers to Odd-Numbered Short-Answer Questions   156
    Multiple-Choice Questions   156
    Answers to Odd-Numbered Multiple-Choice Questions   161

Part XIII Study Guide. Nonparametric Statistics   163
    Part XIII Summary   163
    Learning Objectives   164
    Computational Exercises   164
    Answers to Odd-Numbered Computational Exercises   167
    True/False Questions   169

Answers to Odd-Numbered True/False Questions	170
Short-Answer Questions	170
Answers to Odd-Numbered Short-Answer Questions	171
Multiple-Choice Questions	171
Answers to Odd-Numbered Multiple-Choice Questions	175

Part XIV Study Guide. Effect Size and Power — 177

Part XIV Summary	177
Learning Objectives	179
Computational Exercises	179
Answers to Odd-Numbered Computational Exercises	181
True/False Questions	182
Answers to Odd-Numbered True/False Questions	182
Short-Answer Questions	183
Answers to Odd-Numbered Short-Answer Questions	183
Multiple-Choice Questions	184
Answers to Odd-Numbered Multiple-Choice Questions	187

Part XV Study Guide. Correlation — 189

Part XV Summary	189
Learning Objectives	192
Computational Exercises	193
Answers to Odd-Numbered Computational Exercises	198
True/False Questions	198
Answers to Odd-Numbered True/False Questions	199
Short-Answer Questions	199
Answers to Odd-Numbered Short-Answer Questions	200
Multiple-Choice Questions	201
Answers to Odd-Numbered Multiple-Choice Questions	204

Part XVI Study Guide. Linear Prediction — 205

Part XVI Summary	205
Learning Objectives	206
Computational Exercises	207
Answers to Odd-Numbered Computational Exercises	209
True/False Questions	209
Answers to Odd-Numbered True/False Questions	210
Short-Answer Questions	210
Answers to Odd-Numbered Short-Answer Questions	211
Multiple-Choice Questions	211
Answers to Odd-Numbered Multiple-Choice Questions	214

Part XVII Study Guide. Review — 215

Part XVII Summary	215
Computational Exercises	217
Answers to Odd-Numbered Computational Exercises	222
True/False Questions	222
Answers to Odd-Numbered True/False Questions	223
Short-Answer Questions	223
Answers to Odd-Numbered Short-Answer Questions	224
Multiple-Choice Questions	224
Answers to Odd-Numbered Multiple-Choice Questions	228

Part I Study Guide

Preliminary Information

Part I Summary

Module 1

- Statistics is not a difficult subject, but it can seem that way. One of the difficult aspects of statistics is that it uses many different terms that can be confusing when you are first learning the material. You should become familiar with all the terms as you read the textbook to help you better understand the material. Another aspect that is commonly thought of as difficult is the mathematical component of statistics. However, the actual math that will be used for your introduction to statistics is relatively simple. In fact, in this textbook the hardest mathematical operations are exponents (x^2) and square roots ($\sqrt{\ }$). The difficult part of learning this material will be understanding the logic behind the math. To better grasp this logic, it is important for you to become fluent with the different terms.
- Here is a brief list of common statistical terms and their definitions:
  - *Case/subject:* A unit of study. The term *subject* is used when the cases are humans.
  - *Sample:* A group of subjects in a study that are part of a larger group.
  - *Population:* A large group from which samples are drawn. We use samples to learn about populations.
  - *Statistics:* A number that summarizes a sample.
  - *Parameter:* A number that summarizes a population.
  - *Variable:* A measurement that can vary from person to person in a sample. You can consider the height and weight of five different people to be variables because they will be different (vary) from person to person.
  - *Constant:* A measurement that remains the same for all cases.
  - *Uppercase letters:* These are used to represent variables.
  - *Bar over a letter:* This represents an average.
  - *M:* The mean of a sample.
  - *p:* The probability that an event will occur.
  - *q:* The probability that an event will not occur.
  - *N, n:* The number of cases. N refers to the number of cases in a population, whereas n refers to the number of cases in a sample.
  - *Subscripts:* Reference to a specific case. X_1 is the first case's score on variable X.
  - *Wavy parallel lines:* Indicate an approximation as opposed to =, which indicates a definite amount.
  - *< and >:* Less than a value and more than a value.
  - *Absolute value:* The value of the number, regardless of the sign.
  - *Summation:* Add up all scores for a particular variable.

- *Reciprocal:* Divide one by the number. The reciprocal of 7 would be 1/7.
- *Exponent (superscripted number):* Tells you to multiply a number by itself as many times as the exponent, or superscripted number.
- *Radical sign:* Take the square root of the number under the radical sign.

- Here are a few simple rules to keep in mind as you work through computations:
  - Dividing a number is equivalent to multiplying that number by its reciprocal.
  - When multiplying negative and positive numbers, the product will always be negative. When multiplying two negative numbers or two positive numbers, the product will always be positive.
  - Fractions, decimals, and percentages are different ways to represent the same amount; 1/4 = .25 = 25%.
  - You can expect many of your mathematical operations to have multiple decimal places. It is common to round all work to three decimal places and the final answer to two decimal places.
  - When rounding, values of .5 or greater are rounded up, meaning one is added to the next digit; .56 becomes .6. Alternatively, values of .4 or lower are rounded down, meaning the next digit remains the same; .54 becomes .5.
  - Always remember the order of operations in math. These are (1) any operation in parenthesis, (2) exponents and square roots, (3) multiplication and division, and (4) addition and subtraction. The acronym PEMDAS (aka "Please Excuse My Dear Aunt Sally") can help you to remember this order.
  - Finally, remember that you may need to reorder an equation to solve for an unknown value. Remember the basic rule of algebra that states that what is done to one side of an equation (equal sign) must be done to the other. For example, if $12 = 4x + 4$, subtract 4 from both sides to give $12 - 4 = 4x$. Both sides can then be divided by four, which provides the answer of $x = 2$.

- Σ is the symbol for summation. ΣX indicates that you should sum all the values of X.

Module 2

- Statistics are used to investigate variables. Just as there are different types of variables (age, height, gender, occupation, etc.), there are different scales in which variables can be measured. Measurement refers to the value that is assigned to a specific trait. The meaning that you assign to a certain measurement depends on the scale of measurement that was used. In statistics, there are four types of scales of measurement: nominal, ordinal, interval, and ratio.
- A *nominal scale* is one that is divided into distinct categories. Additionally, the categories are not ranked; one category is not higher or lower than the next. Favorite color would be an example of a nominal scale. My favorite color is blue whereas yours may be red. Personal opinion aside, my preference for blue is no better or worse than your preference for red. Nominal scales are used for statistical tests that focus on comparing distinct groups.
- An *ordinal scale* is one that ranks people in order, but the precise difference between two people is unknown. For example, you might classify one person as attractive and another as gorgeous. You know from these descriptors that gorgeous is above attractive, but you cannot be certain of the exact amount of improvement from one rank to the next.
- An *interval scale* is similar to an ordinal scale in that individuals are ranked in order, but in contrast to an ordinal scale, the precise difference between each individual is

known. For example, the Fahrenheit temperature scale is an interval scale. When comparing 36° and 12°, you can be certain that there is a 24° difference between the two temperatures.
- A *ratio scale* is similar to an interval scale, but possesses a *true zero*. A true zero means that a score of zero represents a true absence of the variable being measured. The amount of money in your wallet at this moment is a ratio scale. If the amount in your wallet is $0, then you have an absence of money.
- Variables can also be considered either continuous or discrete. *Continuous variables* have values that can fall anywhere on the scale, including between two adjacent values. An example of a continuous variable would be time (as specified by minutes). You could have a value 1 min and 3 s. In comparison, *discrete variables* have values that cannot fall between adjacent values. An example of a discrete variable would be the number of students in your class because there cannot be anything less than a full student.
- Continuous variables are defined by real limits. A *real limit* is defined as ±.5 of the scale unit. This makes adjacent scores meet. Thus, for a score of 10, the *upper real limit* would be 10.5 and the *lower real limit* would be 9.5. It is important to note that an observed score will never fall at the real limit.

Learning Objectives

Module 1

- Become familiar with common statistical terms and symbols
- Review common arithmetic rules and functions

Module 2

- Classify data according to their level of measurement
- Distinguish between discrete and continuous scores
- Establish real limits for continuously scored data

Computational Exercises

1. $(3)(6) + 2 =$
2. $\dfrac{(14 + 13)}{3} =$
3. $\dfrac{(3 + 7)2}{4} =$
4. Round your answer to two decimal places: $\dfrac{(9 + 13)6}{8} =$
5. Round the following numbers to two decimal places:
   4.576
   3.2134
   8.9431

6. What is the reciprocal of the following:
    a. 4
    b. 3
    c. 2
    d. 2/3

7. Complete the following chart:

Fraction	Percentage	Decimal
9/45		
	92%	
		.35

8. Complete the following chart:

Fraction	Percentage	Decimal
≈6/9		
	43%	
		.7391

9. Solve the following equations remembering the rules of the order of operations:
    a. $(12+45)-(54-9)=$
    b. $\dfrac{12+56}{4}+\dfrac{12}{2+4}=$
    c. $\dfrac{12+56+12}{4+2+4}=$
    d. $(4+3)^2-(20-7)=$
    e. $\sqrt{5+11}+\left(\dfrac{5+4}{3}\right)^2=$

10. Isolate X in the following equations:
    a. $X+15=25$
    b. $2X+3=9$
    c. $6X-7=41$
    d. $\dfrac{12X+18}{10}=132$
    e. $\dfrac{12-3}{2}=45X$

11. Solve for X in the following equations:
    a. $X+4=9$
    b. $4X-7=2$
    c. $2X-(3+4)^2=83$

12. Solve for X in the following equations:
    a. $aX + b = c$
    b. $\dfrac{aX - b}{c} = Z$
    c. $\dfrac{b(X - a)}{(c + d)} = Z$

13. Expand and solve the following expressions if $X_1 = 2$; $X_2 = 4$; $X_3 = 9$:
    a. $\Sigma(X) =$
    b. $(\Sigma(X))^2 =$
    c. $\Sigma(X^2) =$
    d. $\Sigma(2X) =$
    e. $2\Sigma(X) =$

14. Expand and solve the following expressions if $X_1 = 10$; $X_2 = 5$; $Y_1 = 3$; $Y_2 = 2$
    a. $\Sigma(X + Y) =$
    b. $\Sigma(X^2 + Y^2) =$
    c. $\Sigma(X + Y)^2 =$
    d. $\Sigma(2X + 3Y) =$
    e. $3\Sigma(X + 3Y) =$

15. Expand the following binomials:
    a. $2(a + b) =$
    b. $(a + b)^2 =$
    c. $3(a + b)^2 =$

Answers to Odd-Numbered Computational Exercises

1. 20

3. 5

5.
   a. 4.58
   b. 3.21
   c. 8.94

7.

Fraction	Percentage	Decimal
9/45	20%	.2
23/25	92%	.92
7/20	35%	.35

9.
   a. 12
   b. 19
   c. 8

d. 36
   e. 13

11.
   a. X = 5
   b. X = 2.25
   c. X = 66

13.
   a. 15
   b. 225
   c. 101
   d. 30
   e. 30

15.
   a. $2a + 2b$
   b. $a^2 + 2ab + b^2$
   c. $3a^2 + 6ab + 3b^2$

True/False Questions

1. Statistics are used to summarize populations whereas parameters are used to summarize samples.

2. Both X-bar and *M* are symbols for means.

3. *p* and *q* are related in that as one increases the other decreases.

4. The absolute value of −8 is −8.

5. In doing a complicated math problem, you should complete work in parentheses first.

6. 6.42338139 rounded to two digits is 6.42.

7. $\Sigma(X + Y)$ is the same as $\Sigma X + \Sigma Y$.

8. The scale of measurement for the favorite animals of your classmates would be an ordinal scale.

9. In an interval scale, there is an equal amount of distance between adjacent ranks, and there is a true zero.

10. You are attending a swim competition and notice that the swimmers are ranked by their time. Their ranking alone is an ordinal scale.

11. Height is an interval measure.

12. Flavors of ice cream would be discrete and ordinal scale.

13. The number of trumpet players in an orchestra is a discrete variable.

14. The distance that you can throw a football is a continuous variable.

15. For a scale with a high score of 4 and unit scale of 1, the real limits are 3 and 5.

Answers to Odd-Numbered True/False Questions

1. False
3. True
5. True
7. True
9. False
11. False
13. True
15. False

Short-Answer Questions

1. Describe how a sample and a population differ and how they are related.
2. What is precision in mathematical calculation? How precise should you be when working on a question in statistics? When stating your answer?
3. In a horse race, Johnny Smith places first, while Sandy Jones places fourth. Is Johnny four times as fast as Sandy? Why or why not?
4. What scale of measurement best classifies the following:
    a. Number of calories in a candy bar
    b. Colors in a bouquet of flowers
    c. Elevation in reference to sea level
5. What is the highest scale of measurement for the following:
    a. Favorite TV channels
    b. Number of pets you can own
    c. Responses to a questionnaire that ranges from −5 to 5
6. What is a nominal scale?
7. What are the limitations of a nominal scale?
8. What is the difference between an interval and a ratio scale?
9. What is meant by a ratio scale having an absolute zero?
10. What advantage does an interval scale have in comparison to an ordinal scale?
11. Distinguish between a continuous and a discrete variable.
12. You are interested in obtaining the following information from the class. List the best scale of measurement for each:
    a. Favorite movie
    b. Grade in last math class
    c. Self-rating on a 0 to 10 scale of confidence for statistics

13. You are asked by a professor to develop a grading scheme for your next class. Develop a separate grading scheme for a nominal scale, ordinal scale, interval scale, and ratio scale.

14. Jamal proudly states that he scored twice as high as Rene on an English test and therefore is twice as smart as she is in the subject. What scale must the English test use for Jamal's claim to be true?

15. You are conducting a study on social anxiety. You decide to classify your subjects as either having social anxiety or not having social anxiety. You then give them a question with a rating scale of −7 (very anxious) to 7 (very calm). They are then placed in a social situation and timed to determine how long it takes for them to leave the situation. Identify the four different scales of measurement used in this study and to what variables they correspond.

Answers to Odd-Numbered Short-Answer Questions

1. Samples are small subgroups of populations. Populations consist of larger groups that contain all the individuals in your area of interest. They are related in that samples are derived from populations and are used to infer about populations.

3. He is not four times as fast. You cannot determine that type of information from an ordinal scale of measurement.

5.
   a. Nominal
   b. Ratio
   c. Interval

7. Nominal scales are unable to explain rank or order.

9. An absolute zero indicates a complete lack of the trait being studied.

11. Continuous variables have values that can fall anywhere on the scale. Discrete variables have distinct intervals, and a value cannot fall between the intervals.

13. Answers will vary.

15. Nominal—social anxiety/not socially anxious; Interval—rating scale; Ratio—time in a social encounter; Nominal—male/female

Multiple-Choice Questions

1. If $p = .45$, then $q =$
   a. .45
   b. .65
   c. .55
   d. 1

2. You are interested in studying the mating habits of the remaining 4,000 chimpanzees in the wild. However, you are unable to find all 4,000 chimps and so you observe 200. In this example,
   a. The 4,000 chimpanzees are a population and the 200 are a sample
   b. The 200 chimpanzees are a population and the 4,000 are a sample

c. The 200 chimpanzees are a case and the 4,000 are a subject
d. The 200 chimpanzees are a subject and the 4,000 are a case

3. The average child in your area has a height of 22 in. What is the appropriate symbol to represent this measurement?
   a. N
   b. *n*
   c. M
   d. X

4. What is another way to express 4/5?
   a. .7
   b. 45%
   c. .9
   d. 80%

5. What is another way to express ≈.67?
   a. 4/7
   b. 4/6
   c. 65%
   d. 46%

6. According to the conventions of the social sciences, what is the solution: 4.541 + 6.327?
   a. 10.868
   b. 10.87
   c. 10.9
   d. 10

7. According to the conventions of the social sciences, what is the solution: 9.87532 + 10.78672?
   a. 20.66
   b. 20.66204
   c. 22.66
   d. 19

8. What is another method to express $\Sigma(3X+4Y)$
   a. $(12) \Sigma (X+Y)$
   b. $\Sigma(3X) + \Sigma(4Y)$
   c. $3\Sigma(X) + 4\Sigma(Y)$
   d. $\Sigma(3X) + \Sigma(4Y)$

9. Solve for Z: $15 = \dfrac{Z(12+8)}{5}$
   a. Z = 20
   b. Z = 3.75
   c. Z = 4.25
   d. Z = 10.50

10. Solve for X: $20X - 18 = 6(X + 4)$
    a. X = 4
    b. X = 5.5
    c. X = 12
    d. X = 3

11. If you were interested in measuring the salary of teachers, what is the highest order scale you could use?
    a. Nominal
    b. Ordinal
    c. Interval
    d. Ratio

12. You are interested in studying depression. You obtain a sample of 100 people and divide them into the following categories: Depressed and Not Depressed. These categories are an example of
    a. A discrete variable with an ordinal scale
    b. A continuous variable with an ordinal scale
    c. A discrete variable with a nominal scale
    d. A continuous variable with a nominal scale

13. Which scale would be considered the lowest level of measurement?
    a. Nominal
    b. Ordinal
    c. Interval
    d. Ratio

14. You are interested in measuring the sizes of amoebas. However, the ruler on your microscope has broken and you are no longer able to measure them in a proper unit of measurement. As an alternative, you decide to classify them as small or large. It would be impossible for an amoeba to fall between each category. This scale is an example of
    a. A discrete variable with an ordinal scale
    b. A continuous variable with an ordinal scale
    c. A discrete variable with an interval scale
    d. A continuous variable with an interval scale

15. You obtain a score of 110 on a statistics aptitude test scored in 1-point units. Your upper real limit (UL) and lower real limit (LL) are
    a. UL = 110, LL = 109.5
    b. UL = 110.5, LL = 109
    c. UL = 110.5, LL = 109.5
    d. UL = 111, LL = 109

16. The real limits of a scale that ranges from 1 to 7 in 1-point units are
    a. UL = 1, LL = 7
    b. UL = 1.5, LL = 6.5
    c. UL = 0, LL = 8
    d. UL = .5, LL = 7.5

17. You have created a scale measuring alcohol abuse that uses the following scale: A lot, A little, Not much, None. Your scale is
    a. Ordinal
    b. Interval
    c. Ratio
    d. Could be ordinal or interval as per the debate on measurement.

18. A chef is developing feedback cards to determine if people like his food. He creates a card that provides people with an option to check a box representing that they either enjoyed the food or disliked the food. What type of scale is this?

a. Nominal
b. Ordinal
c. Interval
d. Ratio

19. (Refer to Question 18) The chef is upset that almost 50% of those that completed the cards checked the "disliked" box. He decides to give his customers more options to choose from to get a better understanding of their opinion. His new scale ranges from −5 to +5. What type of scale is this?
    a. Nominal
    b. Ordinal
    c. Ordinal or interval as per debate on measurement
    d. Ratio

20. An ordinal scale
    a. Ranks scores in order on a continuous scale
    b. Places scores in discrete ranked categories
    c. Ranks scores in order and the distance between adjacent ranks is equal
    d. Possesses a true zero point

Answers to Odd-Numbered Multiple-Choice Questions

1. c
3. c
5. b
7. a
9. b
11. d
13. a
15. c
17. d
19. c

Part II Study Guide

Tables and Graphs

Part II Summary

Module 3

- *Frequency tables* present data in an organized manner that enables specific information to be efficiently retrieved. Through the use of a frequency table, you could easily obtain an estimate of how many participants obtained a specific score as well as determine how many participants scored above and below that score. All frequency tables follow a similar format. The left displays all possible values and the right displays the frequency, how many people obtained a specific value. There are three additional columns that you could add to the frequency table to provide you with more information.
- The first additional column creates a cumulative frequency table. A *cumulative frequency table* displays how many scores fall at or below (or possibly above) a specific value. If you were to create a frequency table for test grades, this table would enable you to determine how many students were above your test grade or below your test grade.
- A *relative frequency table* tells you the proportion (percentage) of the total sample that obtained a specific score. Let's say you have a data set with 10 numbers ranging from 1 to 5, and 3 of those numbers are 4s. The relative frequency of 4s would be 30%. This is done by dividing the frequency of a specific score by the number of scores in the data set and then multiplying the quotient by 100. Since relative frequencies are percentages, they must add up to 100%.
- *Cumulative relative frequency tables* provide you with the percentage of scores above or below a specific value. These are created by first finding the relative frequency of each value. Then the relative frequency of the lowest score is added to the relative frequency of the next highest score. Repeat this step until you reach the highest score, which should have a cumulative relative frequency of 100% or 1.00.
- When you have a large number of scores, it can be helpful to group your scores into intervals when creating a frequency table. (Imagine listing all the possible values for the SAT, which has a scale of 0 to 1,600!) When creating a grouped frequency table, all the intervals should be equal in size. There are no standard rules for determining when data should be grouped or the size of each interval.
- Cumulative relative frequencies are sometimes also referred to as *percentile ranks*. Percentile rank indicates the percentage of scores falling at or below a specific score. If you obtain a score of 85 on your next test and your score has a cumulative frequency of .94 percentile rank in your class, you can be certain that you did better than 94% of your class.

- However, since multiple scores can occur at a specified percentile rank (there may have been 6 students with a score of 85), your percentile rank provides only an estimate of your rank. To determine the precise percentile rank, you need to spread that rank across all the subjects with that specific score. This is done by using the UL and LL with the following formula:

$$PR = \left(\frac{\text{cum} f_{LL} + (f_i/i)(X - LL)}{N}\right)100$$

- After using the above formula, assume that the number of scores that fall within the real limits of the score (84.5–85.5) are evenly distributed. To find the precise percentile rank, divide the number of scores at that interval by the proportion (percentage) of scores in that interval and add that amount to the percentage of scores below the interval. If 90% of the scores fell below 84.5 and 4% of the scores fell between 84.5 and 85.5, you can be certain that the true percentile rank of your score was 92%.
- Alternatively, you may be interested in determining the score that corresponds to a specific percentile rank. This can be done using the following formula:

$$P_{PR} = LL + (i/f_i)(\text{cum } f_{UL} - \text{cum } f_{LL})(.5)$$

Module 4

- Although frequency tables provide a neat method for organizing data, graphs can be an even more effective method for presenting information. Information that can be obtained from a graph includes the dispersion, clustering, and location of the majority of scores.
- *Stem-and-leaf displays* are similar to frequency tables, but also provide a visual depiction of the frequency of each score. They are created by listing all data from the highest to the lowest. The left column is the first digit of a score and the right column contains every subsequent digit of all scores that start with the first digit. For very large data sets, you can represent the first digit twice in the left column. The right column should then be divided in half as well so that the top number would display scores with a second digit greater than 5 and the bottom number would display scores with a second digit less than 5.
- When creating a graph, the traditional rules are that the X-axis (abscissa) represents the intervals of the measured variable, and the Y-axis (ordinate) represents the frequency or percentage of scores at each interval. It is important to note that in situations where there is a large frequency of cases for a particular score interval, you can divide the interval on the X-axis.
- There are some rules for creating a graph: (1) The Y-axis should be 3/4 the size of the X-axis; (2) with large data sets, you can collapse intervals on the X-axis so there are at least 5 intervals but not more than 12 intervals; (3) each interval on the X-axis must be equal; (4) the Y-axis must be continuous; and (5) the axes should not stretch or compress the data so that the data are distorted.
- *Histograms* are graphs for continuous data that indicate the frequency of a particular score by bars. The bars are adjacent to one another to indicate that a score could fall between the intervals on the X-axis.
- *Frequency curves* are an alternative to the histogram. A frequency curve is drawn by first creating a histogram and then connecting the midpoint of each bar with a solid line. This provides you with a visual representation of how your data are distributed, as it easily allows you to determine whether the data are clustered around a specific score or if they are spread out, and if they appear heavily lopsided or symmetrical.

- Frequency curves can take on multiple shapes. One of these shapes is the *normal curve*, which appears bell-shaped. This means that it is *symmetrical* in shape, indicating that approximately half the scores fall above the peak (middle score) and half below. The other shapes involve *skew*, which refers to a lopsided distribution. Skew is produced by having a greater concentration of scores at either the upper or lower end of the distribution. If there are a lot of high scores, the distribution is said to be *negatively skewed* as there is a long tail on the left. If there are a lot of low scores, the distribution is said to be *positively skewed* as there is a long tail on the right.
- *Kurtosis* refers to the frequency of scores that are in the middle of the distribution. Distributions with a lot of scores in the center are referred to as *leptokurtic*. Alternatively, distributions with few scores in the center are referred to as *platykurtic*.
- The other two general shapes of the frequency curve are bimodal and rectangular. *Bimodal* distributions have more than one peak. *Rectangular* distributions have uniform responses, which means that the frequency of all the scores is the same.
- When graphing nominal data, it is more appropriate to use a bar graph or a pie graph as these methods express the discrete nature of nominal variables. *Bar graphs* appear similar to histograms except that the bars are separated to indicate that a score could not fall between adjacent categories. *Pie graphs* place the nominal data in a circle with slices to represent the different categories. The size of each slice determines the number of participants in that particular category.

Learning Objectives

Module 3

- Convert scores to frequency—simple, cumulative, relative, relative cumulative, and grouped
- Display scores in frequency tables
- Find percentiles and percentile ranks from tabled data

Module 4

- Distinguish between normally distributed and nonnormally distributed data
- Select the best type of graph for data of a given scale
- Construct various types of graphs from data
- Apply graphing conventions so as not to distort the data

Computational Exercises

Here are the scores of 15 freshman students rating their confidence that they will do well in statistics on a 1 to 10 scale. Use these data for Questions 1 to 8:

5	10	10
7	4	4
3	2	10
2	1	1
2	10	9

1. Arrange the scores into a frequency table in descending order. How many students ranked their confidence as 5? As 6? As 4?

2. Add a column to the table you created for Question 1 to show the cumulative frequency of the scores. How many students ranked their confidence as less than 7? As greater than 4? As less than 10?

3. Add a column to the table you created for Question 1 that shows a relative frequency for each score. What percentage of students ranked their confidence as 3? As 8?

4. Add a column to the table that you created for Question 1 that shows the cumulative relative frequency.

5. What is the percentile rank for a person who rated himself or herself at 5?

6. What is the percentile rank for a person who rated himself or herself at 4?

7. What score falls at the 40th percentile rank?

8. What score falls at the 33rd percentile rank?

NOTE: The following data represent the time for 12 participants to react to a traffic light while under the influence of alcohol in a driving simulation. Use these data for Questions 9 to 13.

$$47, 48, 48, 49, 49, 49, 50, 50, 50, 51, 51, 52$$

9. Create a stem-and-leaf display for this data.

10. Create a histogram that accurately portrays these data using 1-point intervals.

11. Create a frequency curve for these data using 1-point intervals.

12. Describe the shape of this distribution in terms of skewness and kurtosis.

13. If you were to create a positively skewed distribution, where would you need to add scores? A negatively skewed distribution? What would you need to do to make the distribution bimodal? Rectangular?

14. State how you would expect the distributions for the following variables to appear:
    a. Time it takes members of a senior citizens group to run a mile
    b. Amount of money earned in the first year after completing college

15. State how you would expect the distributions for the following variables to appear:
    a. Age of football players
    b. Happiness during a wedding

16. Which would be the appropriate methods for graphing the following data?
    a. Number of trophies won by tennis players
    b. Length of time in therapy for obsessive-compulsive disorder

17. Which would be the appropriate methods for graphing the following data?
    a. Number of words spoken by 3-year-olds
    b. The profit margins of six different companies for last year

Answers to Odd-Numbered Computational Exercises

1.

X	f
10	4
9	1
8	0
7	1
6	0
5	1
4	2
3	1
2	3
1	2

No. of 5s: 1; No. of 6s: 0; No. of 4s: 2

3.

X	f	Cum f	Rel f (%)
10	4	15	26.67
9	1	11	6.67
8	0	10	0.00
7	1	10	6.67
6	0	9	0.00
5	1	9	6.67
4	2	8	13.33
3	1	6	6.67
2	3	5	20.00
1	2	2	13.33

5. 60%

7. A score of 3

9.

Stem	Leaf
5	2
5	1 1
5	0 0 0
4	9 9 9
4	8 8
4	7

11.

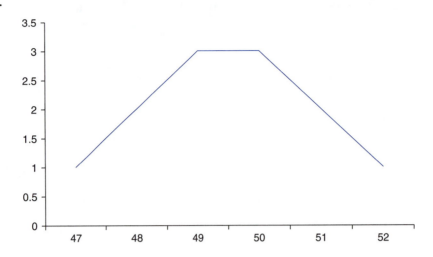

13. To create a positively skewed distribution, you would need to add a large number of scores below 47 or a few scores greater than 52. To create a negatively skewed distribution, you would need to add a large number of scores above 52 or a few scores less than 47. To create a bimodal distribution, you would need to add a second group of scores with a similar peak. To create a rectangular distribution, you would need to make it so that all scores have an equal frequency.

15.
   a. Positively skewed, the majority of players will be younger
   b. Negatively skewed, the majority of people are very happy

17.
   a. Frequency curve, histogram
   b. Pie chart, bar graph

True/False Questions

1. In creating a frequency table, you do not need to list scores that have a frequency of zero.

2. Frequency tables are used to organize information more efficiently.

3. A cumulative frequency of 10 means that there are 10 scores below this particular score.

4. Relative frequencies are the proportion of scores at or below a specific score.

5. In a sample containing $n = 20$ participants, 6 of the participants obtained a score of 12 on a measure. The relative frequency of 12 for this sample is 30%.

6. The sum of all the relative frequencies for a sample must be 100%.

7. The cumulative relative frequency of the lowest score in a data set is always 0%.

8. When creating a grouped frequency table, you should have a minimum of 5 intervals.

9. A score's percentile rank depends on the number of scores there were at that interval and the upper and lower limits of the score.

10. Pie charts and bar charts are excellent methods for graphing continuous data.

11. In a stem-and-leaf display, it can be useful to display each "stem" twice when you have a large amount of data.

12. The height of a bar in a histogram indicates the frequency of a score.

13. A distribution with a lot of high scores and very few low scores would be considered positively skewed.

14. In a symmetrical distribution, the majority of the scores are above the midpoint.

15. Kurtosis refers to the height of the middle scores of a distribution.

Answers to Odd-Numbered True/False Questions

1. False
3. False
5. True
7. False
9. True
11. True
13. False
15. True

Short-Answer Questions

1. What are the advantages of organizing data in a frequency table as opposed to viewing them as raw scores?

2. You are creating a frequency table for a test anxiety scale that ranges from 1 to 10. After administering the test to 25 people, you notice that no one scored a 5 or an 8. How many intervals should your scale have?

3. If you wanted to determine how many scores were in a data set, which frequency table column would provide this information most efficiently?

4. What does a relative frequency tell you?

5. You are checking the grade of your last English test and notice that your professor provided you with a cumulative relative frequency table as well. You notice that cumulative relative frequency of your score was .59. What does this mean?

6. When should you consider making a grouped frequency table as opposed to a regular frequency table?

7. You are trying to arrange the SAT scores of your high school in a frequency table. You realize that you need to use a grouped frequency table because of the large scale (0–1,600) and so you decide to create intervals of 400. Was this a good idea?

8. What is a percentile rank?

9. You are investigating a measure of well-being in a geriatric population. The scale has a range of 0 to 50. The cumulative relative frequency for a score of 44 is 34% and the relative frequency is 4% for that interval. Why is the percentile rank of those with a score of 44 equal to 32% and not 34%?

10. What can you do to better organize a stem-and-leaf display if you have a large number of cases in a data set?

11. Why are the bars on a histogram connected?

12. What is the difference between a positively skewed distribution and a negatively skewed distribution?

13. You notice that in a recent survey for a new TV show, the scores have a rectangular distribution. What does this mean about the responses in your sample?

14. What is kurtosis and what are its two forms?

15. What do the slices on a pie graph represent?

Answers to Odd-Numbered Short-Answer Questions

1. The data are organized in an efficient manner such that you can easily determine how many of any particular score are in the data set. When viewing raw scores, especially those out of order, it can be difficult to determine this information.

3. The cumulative frequency column.

5. Fifty-nine percent of the class scored the same or lower than you and 41% of the class scored higher.

7. This was probably not a good idea. You would have 4 groups and you would expect not many people to fall in the lowest group. Also, you would lose a great deal of precision as most people would score between 800 and 1,600, which would be just two intervals.

9. Percentile ranks are based on the real limits of a score. The cumulative relative frequency of 34% with a relative frequency of 4% indicates that there were 30% of the scores below this interval and 4% of the scores in this interval. Also, you can assume that the scores were evenly distributed within this interval. Thus, you have to add the number of scores below (30%) to 1/2 of the number of scores within (2%). This provides you with a percentile rank of 32%.

11. This indicates that the values could fall anywhere between the intervals.

13. Each score (interval) on your survey received an equal number of responses.

15. The proportion of cases that fell in a particular category. This is used for nominal data.

Multiple-Choice Questions

NOTE: The following scores represent the amount of fear (15–1) experienced when going through a haunted house on Halloween. Use the following table for Questions 1 to 7.

Part II Study Guide

Score	Frequency
15	2
14	5
13	8
12	9
11	12
10	15
9	15
8	16
7	18
6	20
5	21
4	23
3	24
2	15
1	10

1. What is the cumulative frequency of participants who ranked their fear as 5 or lower?
   a. 21
   b. 72
   c. 93
   d. 103

2. How many people visited the haunted house this past Halloween?
   a. 123
   b. 213
   c. 15
   d. 120

3. What is the approximate relative frequency of people who provided a rating of 10?
   a. .07
   b. .70
   c. .05
   d. .50

4. What is the cumulative relative frequency of those who provided a rating of 15?
   a. .10
   b. .4
   c. .98
   d. 1.0

5. If you were to draw a relative frequency curve of these data, how would you describe the shape of this distribution?
   a. Symmetrical
   b. Positively skewed
   c. Negatively skewed
   d. Bimodal

6. What is the percentile rank of a score of 3?
   a. 23.23%
   b. 23.00%

c. 15.00%
d. 12.68%

7. What is the cumulative relative frequency of a score of 11?
   a. 27.45%
   b. 88.73%
   c. 94.31%
   d. 74.85%

8. You are interested in determining how moviegoers rated a new film on the opening day on a scale of 1 to 8. After collecting data, you discover that a large proportion of the sample provided ratings of 4 and 5. How would you expect this distribution to appear?
   a. Positively skewed
   b. Negatively skewed
   c. Leptokurtic
   d. Platykurtic

9. You are working at an advertising firm and you want to determine the interest of a focus group in a new product. If the scale ranges from 0 (no interest) to 10 (great interest), what type of distribution would be most preferable?
   a. Positive skew
   b. Negative skew
   c. Bimodal
   d. Symmetrical

10. A professor announces to his class that a large portion of the class did really well on the last test. However, another large portion appeared to do poorly. How should the professor expect the distribution to look?
    a. Positive skew
    b. Negative skew
    c. Bimodal
    d. Symmetrical

11. The X-axis is sometimes referred to as the
    a. Abscissa
    b. Ordinate
    c. Platykurtic
    d. Leptokurtic

12. Histograms are best used for
    a. Dichotomous variables
    b. Continuous variables
    c. Discrete variables
    d. Nominal variables

13. In a study examining the rate at which a person can mentally rotate an object, the results indicate that the majority of people take 15 s with an equal number of participants falling above and below this time. You could expect the relative frequency curve of these scores to appear
    a. Symmetric
    b. Positively skewed
    c. Negatively skewed
    d. Bimodal

14. After giving his class a quiz worth 10 points (1–10 scale), your professor notices that the distribution was positively skewed. What conclusion should the professor draw from this information?
    a. The test was very hard
    b. The test was too easy
    c. The test was appropriate
    d. The majority of the students did well

15. What would be the best method to graph data obtained for the question "What is your favorite type of pie?"
    a. Histogram
    b. Relative frequency curve
    c. Bar graph
    d. Stem-and-leaf display

16. In a study of post-traumatic stress, you obtain ratings of the number of flashbacks a person has in a month. If you have a sample of 75 individuals, and 40 have indicated they have had flashbacks twice in the last month, what is the relative frequency of those with two flashbacks in the last month?
    a. .63
    b. .40
    c. .65
    d. .53

17. Bar and pie charts are best used for which type of variables?
    a. Continuous
    b. Ordinal
    c. Nominal
    d. Ratio

18. You administer a test of reading comprehension having a scale of 1 to 20 to a group of 40 six-year-olds. Precisely two participants obtain each possible score. How will this distribution appear?
    a. Rectangular
    b. Normal
    c. Positively skewed
    d. Negatively skewed

19. You had a percentile rank of 73 in your high school class. What percentile of students were ranked above you?
    a. 73
    b. 27
    c. 50
    d. 23

20. A common symptom of depression is a lack of desire to do things that are entertaining. If you were to ask a sample of 100 individuals with depression about how many fun activities they have done in the past week, how would you expect the distribution to appear?
    a. Rectangular
    b. Symmetrical
    c. Positively skewed
    d. Negatively skewed

Answers to Odd-Numbered Multiple-Choice Questions

1. c
3. a
5. b
7. b
9. b
11. a
13. a
15. c
17. c
19. b

Part III Study Guide

Central Tendency

Part III Summary

Module 5

- *Central tendency* provides a single value that best describes, or is most representative of, the entire set of scores.
- The *mode* is the most commonly occurring score in the data set. Although the mode is referred to as a measure of central tendency, it does not necessarily occur at the center of the data set (there may not be an equal number of scores above or below the mode). The mode is the least stable measure of central tendency, meaning that it may change drastically from sample to sample.
- The *median* is the center score; half of the scores in the distribution are above the median and half of the scores in the distribution are below the median. In other words, the median is the score that occurs at the 50th percentile. The median does not have to be an actual score. For example, the mean of the distribution 3, 4, 6, 7 would be 5. You can use the following formula for a precise measure of the median:

$$Mdn = LL + (i)\left(\frac{.5n - \text{cum } f_{\text{below}}}{f}\right)$$

- The *mean* is the average score for a data set and is symbolized as M for samples and as μ for populations. The mean is the most commonly used and most stable measure of central tendency. The formula for a mean is as follows:

$$M = \frac{\sum X}{N}$$

- There are three important aspects to the mean. First, the distance of all scores above the mean is equal to the distance of all scores below the mean. Second, the mean includes all values of the data in its calculation, which indicates that each score in the distribution matters. Finally, the mean is also a sensitive measure of central tendency in that a change in any score in the data set will change the mean.
- An extreme score in a data set, one that is drastically different from the others, is called an *outlier*. Outliers can influence the measure of central tendency that is used, as the median may be a more appropriate measure if there are many outliers in a data set.

- The skew of a distribution will affect the location of the measures of central tendency. In a symmetrical distribution, the mean, median, and mode are all equal. However, in a skewed distribution, the values for the measures of central tendency are as follows. Positively skewed distributions: mode < median < mean. Negatively skewed distributions: mode > median > mean.
- It is not appropriate to report any measure of central tendency when you have multimodal information. A graph is the best method for displaying this distribution.

Learning Objectives

Module 5

- Calculate various measures of central tendency—mode, median, and mean
- Select the appropriate measure of central tendency for data of a given measurement scale and distribution shape
- Know the special characteristics of the mean that make it useful for further statistical calculation

Computational Exercises

NOTE: The following are the ratings of newlyweds on a marital satisfaction inventory. The scale ranges from 1 to 10. Use this data for Questions 1 to 7.

6, 2, 3, 4, 5, 2, 6, 3, 4, 5, 6, 8, 7, 7, 5, 6, 8, 7, 5, 3, 4, 4, 5

1. What is the mode amount of marital satisfaction?
2. What is the median amount of marital satisfaction?
3. What is the numerical distance of the scores above the mean? What is the numerical distance of the scores below the mean? Which is greater?
4. Based on the measures of central tendency (use a ballpark estimate for the median), how would we expect the distribution of these scores to appear? (Do not sketch your answer.)
5. If we were to add a score of 45 to this distribution, what would the new measures of central tendency be? Which measure would have been the most influenced?
6. If we were to add another score of 5 to the original distribution, what would the new measures of central tendency be? Which score would be the most influenced?
7. In Question 5, we added a score of 45 to the distribution. How would we expect the shape of the distribution to change if this score were added? How would we expect the shape to look if we were to add another score of 5?

NOTE: The following are data obtained from tourists rating their experience at a particular hotel during a 1-week stay in Hawaii. The scale that was used ranged from 1 to 50. The $M = 29$. Use the following data to answer Questions 8 to 12.

Score	Frequency
50	9
40	15
30	12
20	8
10	6

8. What is the modal rating?

9. What is the median rating?

10. How would you expect the distribution to appear based on the measures of central tendency? (Do not sketch the distribution.)

11. Which measure of central tendency would best describe the distribution?

12. If we were to obtain another sample of 50 participants and each of them provided the exact same responses as the ones we had obtained from the first sample (so all frequencies are now doubled), what would the measures of central tendency become? Which was most affected by this change?

13. In a study looking at the distance people commute to work, the average distance was 40 miles. If the sum of all the distances in the sample was 200, what was n for the study?

NOTE: Here are the results of a pop quiz given in an American history course. The quiz was graded on a scale of 1 to 10.

$$2, 1, 9, 3, 2, 4, 5, 6, 7, 8, 2, 3, 7, 7, 2, 4, 7, 8, 5$$

14. What are the mean, median, and mode of these scores?

15. Which measure of central tendency would best describe this distribution?

Answers to Odd-Numbered Computational Exercises

1. Mean = 5, mode = 5

3. Sum above = 16. Sum below = −16. The difference is 0. Neither is different. The mean is the balance point of the distribution.

5. Mean = 6.67, median = 6.1, and mode = 5. The mean has changed the most.

7. If we were to add a score of 45, the distribution would have a strong positive skew. If were to add another score of 5, the distribution would still appear normal (symmetrical).

9. $25 + (10)\left(\dfrac{.5(50) - 14}{12}\right) = 34.17$

11. The median because it is a skewed distribution.

13. $n = 5$

15. The distribution is bimodal and so there is no appropriate measure of central tendency.

True/False Questions

1. Measures of central tendency provide a single value that is considered highly representative of a distribution.
2. The mean is always the optimal measure of central tendency.
3. Each measure of central tendency (mean, median, and mode) can be found in the exact center of all distributions.
4. The mode is a highly stable measure of central tendency.
5. The number of scores falling above and below the median are always equal.
6. The median is the score that occurs at the 50th percentile rank.
7. One of the reasons that the mean is considered a good measure of central tendency is because it includes every score in its computation.
8. The total distance of the scores below the mean will always be equal to the total distance of scores above the mean.
9. In a positively skewed distribution, you can expect all three measures of central tendency to be equal.
10. There is no single best measure of central tendency to describe a bimodal distribution.
11. Outliers are scores close to the mean.
12. Multiple outliers of different values will heavily influence the mode.
13. The median is considered to be a better measure of central tendency in a skewed distribution.
14. When reporting the mode for bimodal data, you should report both modes.
15. In a negatively skewed distribution you can expect the mean > median > mode.

Answers to Odd-Numbered True/False Questions

1. True
3. False
5. True
7. True
9. False
11. False
13. True
15. False

Short-Answer Questions

1. Why are measures of central tendency useful?
2. What does it mean that a score is "representative of an entire data set"?

3. What is the mode? Is the mode a very stable measure of central tendency? Why or why not?

4. Define the median. How sensitive is the median to the changes in the data set?

5. How is the mean different from the median and mode as a measure of central tendency?

6. Sometimes the mean is said to be the center of a seesaw. What does this mean?

7. What makes the mean the measure of central tendency that is the most sensitive to change?

8. What is an outlier? Describe a situation in which an outlier would have a heavy influence on the mean. Which measure of central tendency is the most appropriate to use in this situation?

9. Compare and contrast the locations of the mean, median, and mode in the following distributions: positively skewed, negatively skewed, and symmetrical.

10. You have just completed a study and discover that you have a skewed distribution. Explain which measure of central tendency is the most appropriate to use with skewed data.

11. You are concerned that children are watching an excessive amount of violent TV. You collect a sample and notice that the distribution is negatively skewed. If you wanted to make the biggest impact in the news by reporting the largest value, which measure of central tendency should you report?

12. A cupcake company has asked you to summarize their customers' opinions on a new type of cupcake filling. They have provided their opinions on a scale of 1 to 5. If the data are normally distributed, which measure of central tendency should you report?

13. In looking over the data from the previous study, you notice that the data are symmetrical, but leptokurtic. How would this affect the mean, median, and mode (you may want to sketch the curve for help)?

14. An instructor has just completed teaching a computer science course for the first time, and many of his students did not do well. The distribution of final grades was positively skewed. Which measure of central tendency would be most appropriate to report to the administration for the class?

15. An instructor has only 10 students enrolled in her course. After the first test, she obtains the following grades (on a 0–100 scale): 45, 64, 57, 34, 65, 49, 44, 58, 58, 62. What is the mean score of her class? She wants to help out her students and so she agrees to add 20 points to everyone's grade. How would this change the mean? (*Hint*: You shouldn't need to do more calculations.)

Answers to Odd-Numbered Short-Answer Questions

1. They provide a single score to represent an entire set of numbers. It enables you to have a quick "snapshot" of where the majority of the scores are located.

3. The mode is the most commonly occurring score. It is not a very stable measure of central tendency as the addition of a few scores can drastically change the location of the mode. For example, a data set may have a mode of 3. However, after adding a few more scores, the mode may become 15.

5. The mean differs from the median and mode in that it includes every score in its calculation. This makes it far more sensitive to changes in the data set than the other measures of central tendency.

7. The mean includes all the scores in the distribution in its calculation. If you were to add any score to a data set, the mean would always change, whereas this may not be the case with other measures of central tendency.

9. In a positively skewed distribution, you can expect the mode to be the smallest value, followed by the median, and the mean would be the largest value. In a negatively skewed distribution, the mean would be the smallest value, followed by the median and finally the mode. In a symmetrical distribution, all three measures are equal.

11. You should report the mode because that will be the largest value.

13. This should have no influence on the measures of central tendency because the data are normally distributed.

15. The mean would be 53.6. By adding 20 points to every grade, the mean would also increase by 20 points. The new mean would be 73.6.

Multiple-Choice Questions

1. You are interested in using the most *stable* measure of central tendency. Which measure should you use?
   a. Mean
   b. Median
   c. Mode
   d. All are equally stable

2. How should you report central tendency when you have a bimodal distribution?
   a. Report the mean
   b. Report one mode
   c. Report the median
   d. Report two modes

3. How many scores should you expect to fall above the mode?
   a. 50% of the scores
   b. The mode should be the center such that the numerical distance of values above the mode is equal to the numerical distance of scores below the mode
   c. Less than half the scores
   d. There is no set number of scores you should expect to fall above the mode

4. In the following set of scores, what is the median? 4, 5, 1, 3, 5, 3, 2, 6, 1, 9, 0, 9
   a. 9
   b. 1
   c. 3
   d. 4

5. In the following set of scores, what is the median? 3, 1, 3, 2, 4, 5, 9, 7
   a. 3.5
   b. 4.5
   c. 3
   d. 4

6. What is the difference between M and μ?
   a. M is for population means, and μ is for sample means
   b. M is for sample means, and μ is for population means
   c. M and μ have the same definition
   d. M is for median, and μ is for mean

7. A high school football team is having tryouts and asks 5 participants to throw a football. Here is how far each person threw the ball (in feet): 79, 84, 93, 165, 88. Which distance would be considered an outlier?
   a. 79
   b. 84
   c. 93
   d. 165

8. Using the information from Question 7, the mean of the distances with the outlier would be _____, whereas the mean of the distances without the outlier would be _____.
   a. 101 and 86
   b. 86 and 101
   c. 73 and 132
   d. 132 and 73

9. You are collecting data for a study on hospital service. You have obtained data from patients at 30 different hospitals that have completed a satisfaction survey. The measures of central tendency are as follows: mean = 50; median = 50; mode = 50. What can you tell about all the scores that you have collected?
   a. There are numerous outliers
   b. The scale ranges from 1 to 50
   c. The scores are distributed in a symmetrical shape
   d. All participants provided identical information

NOTE: The following information is the record sales of a music label's nine recording artists in thousands. Use this information for Questions 10 to 15.

$$6, 8, 15, 10, 23, 17, 23, 32, 19$$

10. What is the average number of records (in thousands) sold by a recording artist at this label?
    a. 16.5
    b. 22
    c. 17
    d. 18.9

11. What is the modal number of record sales (in thousands) by an artist?
    a. 17
    b. 27
    c. 15
    d. 23

12. What is the median number of record sales (in thousands) by an artist?
    a. 17
    b. 19

c. 34
d. 23

13. How would you expect this distribution to appear (without sketching it)?
    a. Symmetrical
    b. Positively skewed
    c. Negatively skewed
    d. Bimodal

14. If you were to add one outlier to this distribution what would you expect to happen?
    a. The mean would change but the mode and median would be relatively unaffected
    b. The median would change but the mode and mean would be relatively unaffected
    c. The mode would change but the median and mean would be relatively unaffected
    d. Nothing would change

15. The owner of the record label doesn't like the shape of the distribution of recent sales. If you were to adjust the distribution so that it appeared symmetrical, where would you tell the owner to add scores?
    a. Above the mean
    b. Below the mean
    c. Precisely at the mean
    d. Obtain a few extreme scores above the mean

16. Here is the commute time (hours) of seven people who work at the same company: 4.00, 1.00, 0.50, 0.75, 2.00, 0.25, 1.50. What is the average amount of time it takes for these people to commute?
    a. 2.1
    b. 1.67
    c. 1.43
    d. 1.54

17. Using the information from Question 16, the person with the 4-hr commute begins to work for a different company and is no longer included in the calculation. What is the new mean amount of time it takes for the remaining people to commute?
    a. 1.45
    b. 1.00
    c. 1.47
    d. 0.86

18. (Use information from Questions 16 and 17.) To replace the person who left in Question 17, the company hires a new person who commutes only .1 hr to work each day (they live very close). What is the new mean amount of time it takes for the workers to commute?
    a. 0.87
    b. 1.00
    c. 1.02
    d. 1.10

19. Using the original set of scores (from Question 16), which measure of central tendency would best describe the distribution?
    a. Mean
    b. Median
    c. Mode
    d. None

20. A farmer is interested in finding out what vegetable is preferred by elementary school children. He asks them to select from the following categories: peas, broccoli, carrots, onions, and celery. What measure of central tendency should he use to summarize his data?
    a. Mean
    b. Median
    c. Mode
    d. None

Answers to Odd-Numbered Multiple-Choice Questions

1. b
3. d
5. a
7. d
9. c
11. d
13. c
15. b
17. b
19. b

Part IV Study Guide

Dispersion

Part IV Summary

Module 6

- *Dispersion* is the extent that scores in a distribution are spread out or clustered together. Similar to measures of central tendency, measures of dispersion, or variability, are expressed as a single value.
- The *range* is the difference between the highest score and the lowest score in a data set. The range is the simplest measure of dispersion, and it is very insensitive to changes in the distribution. Adding scores to the center of the distribution will not affect the range. The only way the range can be impacted is by changing the most extreme scores.
- *Variance* (s^2 for samples; σ^2 for populations) is the average squared distance of each score from the mean. The formula for the variance is

$$s^2 = \frac{\sum (X-M)^2}{n}$$

- The definition of the variance becomes more understandable by reviewing the formula. The numerator states that you should obtain the *deviation score* for every score in the distribution. This is found by subtracting each score from the mean. However, all the deviation scores sum to zero (this is a good way to check your work when finding variances). To proceed with obtaining the variance, you must square each deviation score. This will remove any signs (+/−) from the deviation scores and allow them to sum to a value other than zero. You then divide the squared deviation scores by the number of scores in the sample to obtain the variance or the *average squared distance from the mean.*
- Unfortunately, the average squared distance from the mean is difficult to interpret. The variance tells you the average distance from the mean in area units (which we are unable to interpret) as opposed to linear units (which we normally use). *Linear* distance is distance in original units or regular score points (i.e., the linear distance between 3 and 5 is 2). To revert the variance (in area units) back to linear units, you take the square root of the variance. This is referred to as the standard deviation. The *standard deviation* (s for samples; σ for populations) is the average *linear* distance of scores from the mean. The formula for standard deviation is

$$s = \sqrt{\frac{\sum (X-M)^2}{n}}$$

- The standard deviation is a *standardized* measure of dispersion, indicating that it enables you to use it when working with a specific type of distribution called the normal curve. The normal curve will be discussed at length in later chapters.
- It is important to note that not all measures of linear dispersion are standardized. An unstandardized measure of dispersion is the *average absolute deviation* that is similar to the formula for variance but uses the absolute value of each deviation score rather than the squaring technique. Although this provides a more intuitive measure of dispersion, it cannot be used with the normal curve, which limits its use.
- When using a measure of dispersion to describe a sample or population, the statistic or parameter is called a *descriptive statistic*. Alternatively, you can use sample statistics to make guesses about larger populations. This is commonly referred to as *inferential statistics*. You are using the sample data to make inferences (educated guesses) about the larger population.
- When using sample variances and standard deviations to infer about a larger population, it is important to note that your estimate will not be precise. In other words, your sample variance will not equal the population variance. In fact, you can be certain that the sample variance will be less than the population variance. To adjust for this bias, $n - 1$ can be placed in the denominators of the variance and standard deviation formulas when estimating the population standard deviation from a sample. This adjustment will increase the variance (or standard deviation), which will help to better approximate the population variance (or standard deviation).

Learning Objectives

Module 6

- Calculate various measures of dispersion—range, variance, and standard deviation
- Select the appropriate measure of dispersion for data of a given distribution shape and for a given purpose
- Know the special characteristics of the standard deviation that make it useful for further statistic calculations
- Distinguish between descriptive and inferential formulas for the variance and standard deviation

Computational Exercises

The following are test grades for students in a European history class after the first test.

87	89	99	99
98	93	67	89
90	93	70	87
65	94	82	95
70	79	66	90

1. Find the deviation score for each test grade. What is the sum of these deviations?

2. Find the variance for the grades. Find the standard deviation for the grades.

3. How many standard deviations from the mean is the person with the highest grade? The person with the lowest score?

Part IV Study Guide

4. One of those who received a 99 is an exceptional student and turned in an extra credit project that was worth an additional 10 points on the test. What would the variance and standard deviation be with this person's grade being 10 points higher? (Recalculate the measures of dispersion using this new high score.)

5. Reflecting back on your response to Question 4, which of these measures of dispersion changed most?

6. The majority of students in the class got a particular question wrong. As a result, your professor decides to give everyone 2 points for that question. What happens to the measures of dispersion when you have added 2 points to all the original grades?

7. You are curious about how the measures of dispersion would have been changed had the professor doubled everyone's grade. What would the measures of dispersion be if every score is multiplied by 2?

8. How many of the original grades are between 1 and 2 standard deviations above the mean?

9. A researcher is interested in determining how many food pellets rats on a new diet drug have eaten. They discover the average amount of food eaten is 4.5 pellets and the standard deviation 0.75. What is the variance of food pellets?

Questions 10 to 15 are based on the following information: The manager of a shoe store is interested in how many of each shoe size they sold in the last hour of the day. They have made 10 sales; here are the shoe sizes for each person who purchased shoes during that last hour.

7	12
11	8
13	8
7	6
6	12

10. Find the variance and standard deviation for these shoe sizes.

11. At the last few minutes of the hour, three people run in stating that they are in a shoe emergency and ask to purchase shoes. The sizes of these three new people are 8, 6, and 10. How will including these data affect the standard deviation of the scores?

12. How many of these 13 (including those added from Question 11) people fell within 1 standard deviation of the mean?

13. What is the average absolute deviation of the original shoe sizes? How does this compare with the standard deviation?

14. If you were to estimate the population variance of shoe sizes of those that shop at this store from this sample, what formula would you use (use the original 10 scores)? What would the population variance estimated from this sample be?

15. The store manager collects similar data for the following day and finds that the mean is slightly higher, but the standard deviation is now much higher. Which mean is more representative of the average shoe size of those who shop at the end of the day at this store?

Answers to Odd-Numbered Computational Exercises

1.

X	X − M
87	1.9
98	12.9
90	4.9
65	−20.1
70	−15.1
89	3.9
93	7.9
93	7.9
94	8.9
79	−6.1
99	13.9
67	−18.1
70	−15.1
82	−3.1
66	−19.1
99	13.9
89	3.9
87	1.9
95	9.9
90	4.9

3. The highest grade is approximately 1.20 standard deviations above the mean. The lowest grade is 1.7 standard deviations below the mean.

5. The variance is most affected by this change.

7. The standard deviation would be multiplied by 2 as well (22.54). The variance would be multiplied by 4 (507.96).

9. The variance is .56.

11. The new standard deviation would be 2.42.

13. Average absolute deviation = 2.18. It is slightly smaller than the standard deviation.

15. The first mean is more representative because the scores are more tightly clustered about the mean.

True/False Questions

1. The range is a very sensitive measure of central tendency.

2. Adding 10 different scores to the center of a data set will not affect the range.

3. Deviation scores provide a measure of a score's distance from the mean.

4. The sign (+/−) of a deviation score indicates its location with regard to the mean.

5. Deviations scores always sum to 1.

6. The variance is the average distance of a score from the mean in squared deviation units.
7. Standard deviations from different samples of the same measure can be compared.
8. The standard deviation is found from the square root of the variance to revert the measure to linear units.
9. The standard deviation is the most commonly used measure of central tendency because of its interpretability and applicability to the normal curve.
10. The average absolute deviation is used more frequently than the standard deviation.
11. Descriptive statistics are used to describe the characteristics of a sample.
12. The mean and standard deviation are examples of inferential statistics.
13. When using a sample to infer about a population, you should use $n - 1$ in the denominator.
14. Using $n - 1$ in the denominator corrects for the bias of sample variance being larger than the population variance.
15. There is a debate in the social sciences regarding the appropriate use of N as opposed to $n - 1$ in the denominator of strictly sample variances.

Answers to Odd-Numbered True/False Questions

1. False
3. True
5. False
7. True
9. True
11. True
13. True
15. True

Short-Answer Questions

1. Why are measures of dispersion important when describing a sample?
2. What aspects of the range make it a poor measure of dispersion?
3. What are the steps involved in finding the variance?
4. Why is it necessary to square the deviation scores when finding the variance?
5. What unit of measurement is used for the variance? What unit of measurement is used for the standard deviation? How are these different?
6. How are area units changed to linear units?
7. Why can we always expect the deviation scores to sum to zero?

8. In a high school gym class, the teacher is interested in assessing the average and standard deviation of the amount of time it takes for the students run a mile. The teacher notices that one student is exceptionally quick and completes the mile at 4 standard deviations above the mean. Would you consider this person an outlier? Why or why not?

9. How would all three measures of dispersion (variance, range, and standard deviation) be affected by adding a large number of scores close to the mean of any given distribution?

10. What does it mean to be a standardized measure?

11. What is the difference between a standard deviation and an average absolute deviation?

12. Why is the average absolute deviation not commonly used?

13. What are descriptive statistics? What are examples of this type of statistic?

14. What are inferential statistics? Why are they important in our study of statistics?

15. Why do we use $n - 1$ in the denominator of variance when estimating a population variance from a sample?

Answers to Odd-Numbered Short-Answer Questions

1. Dispersion is important because it indicates the extent that scores cluster around the mean or are very distant from the mean. This will help you to determine how viable the mean is as a single descriptor of the data set. For example, a 0 to 10 scale with a mean of 5 and a standard deviation of 1 indicates that on average, a score will deviate 1 point from the mean. This suggests that the majority of scores will fall between 4 and 6, making the mean a very good descriptor. However, if the mean was 5 and the standard deviation was 5, that means the average deviation from the mean was 5 points, indicating that most scores fall anywhere on the scale.

3. First, the deviation scores must be found. Then these scores must be squared. The squared deviation scores are then summed. Finally, this summed squared deviation score is divided by n.

5. Variance is in area units. The standard deviation is in linear units. The difference is that the area units are squared, meaning they are not able to be directly applied to the original scale of measurement. The linear units are on the same metric as the original scale.

7. Deviation scores always sum to zero because they represent each score's numerical distance from the mean. Since the mean is the numerical center of the distribution, the distance of scores above the mean will always equal the distance of scores below the mean.

9. The range would not be affected at all by adding scores to the center of the distribution. However, the variance and range would become smaller.

11. A standard deviation represents the standardized average deviation which is applicable to the normal curve. The average absolute deviation uses the absolute value of the deviation scores, indicating it cannot be used with the normal curve.

13. Descriptive statistics are those that summarize the data in a sample. Examples of this type of statistic are measures of central tendency and dispersion.

15. $n - 1$ is used in the denominator to correct for a sample's underestimation of the variability in a population.

Multiple-Choice Questions

The coach of a basketball team is interested in assessing how well his team shoots free throws. Here is the number of baskets that each member of his team made during their last free throw practice session. Use this information for Questions 1 to 5.

10	0
3	10
11	3
1	1
0	1
7	8

1. The coach is asked to provide a quick measure of the dispersion for his team's free throws. What is the range for his team?
   a. 0
   b. 1
   c. 11
   d. 12

2. The coach has a little more time on his hands and wants to determine the variance. What is the variance for his team?
   a. ≈12.32
   b. ≈16.91
   c. ≈13.21
   d. ≈4.56

3. The coach realizes that the variance is difficult to interpret and now wants to know the standard deviation for his team. What is the team's standard deviation?
   a. ≈5.32
   b. ≈1.23
   c. ≈16.91
   d. ≈4.11

4. Shortly after obtaining these data, a new player is added to the team who makes 10 baskets on free throws. If you were to incorporate this new player's score, which measure of dispersion would not be affected?
   a. Variance
   b. Standard deviation
   c. Range
   d. Mean

5. The coach is interested in using the data from his team to learn about the number of free throw baskets made by all the teams in the league. This is an example of
   a. Descriptive statistics
   b. Inferential statistics
   c. Central tendency
   d. Dispersion

6. A distribution of scores is discovered to be highly leptokurtic. How much dispersion would you expect?
   a. A small amount because many scores are close to the mean
   b. A large amount because many scores are close to the mean

c. A small amount because many scores are distant from the mean
d. A large amount because many scores are distant from the mean

7. How much dispersion can you expect with a constant?
   a. A great deal of variability
   b. A moderate amount of variability
   c. Minimal variability
   d. No variability

8. You want to find the standard deviation for a set of scores to describe the sample to someone else. Which formula should you use?
   a. $s^2 = \dfrac{\sum (X-M)^2}{n}$
   b. $s = \sqrt{\dfrac{\sum (X-M)^2}{n}}$
   c. $s^2 = \dfrac{\sum (X-M)^2}{n-1}$
   d. $s = \sqrt{\dfrac{\sum (X-M)^2}{n-1}}$

Here are the responses of 10 individuals on a survey assessing satisfaction at the local Department of Motor Vehicles (DMV). The scale is 1 to 10. Use this information for Questions 9 to 14.

5	3
5	6
7	6
1	1
1	6

9. What is the range of these ratings?
   a. 1
   b. 7
   c. 6
   d. 5

10. What is the variance of these ratings?
    a. 6.78
    b. 5.09
    c. 7.89
    d. 4.21

11. What is the standard deviation of these ratings?
    a. ≈2.27
    b. ≈3.47
    c. ≈5.7
    d. ≈6.87

12. What is the average absolute deviation of these ratings?
    a. ≈2.08
    b. ≈3.4

c. ≈2.27
d. ≈4.58

13. How many scores are more than 1 standard deviation away from the mean?
    a. 0
    b. 1
    c. 2
    d. 3

14. What would the standard deviation be if the supervisor of the DMV wanted to make an estimate about the satisfaction of those that have used the DMV?
    a. ≈2.54
    b. ≈1.52
    c. ≈1.1
    d. ≈7.5

15. Late at night, Javier is working on some marketing data that are symmetrically distributed. He is very tired and accidentally substitutes the median instead of the mean when calculating his deviation scores. How will this affect his standard deviation?
    a. It will increase it
    b. It will decrease it
    c. It will not affect it
    d. You need more information

16. In general, how would adding outliers to a distribution affect dispersion?
    a. It would increase dispersion
    b. It would decrease dispersion
    c. Dispersion is unaffected by outliers
    d. It depends on how far the outlier is from the mean

17. You are interested in studying the eye color of your fellow classmates. You obtain the following data: blue, blue, brown, brown, brown, brown, green, green, gray. What would be the standard deviation for this distribution?
    a. 1
    b. 2
    c. 3
    d. You can't calculate a standard deviation for nominal data

Mr. Jones is having difficulty sleeping. He decides to chart the number of hours he has slept over the past five days. Use this information for Questions 18 to 20.

$$\begin{matrix} 4 \\ 3 \\ 7 \\ 8 \\ 2 \end{matrix}$$

18. What is the variance of the number of hours he has slept?
    a. ≈48.45
    b. ≈5.36
    c. ≈7.45
    d. ≈6.89

19. What is the standard deviation of the number of hours he has slept?
    a. ≈2.32
    b. ≈3.45
    c. ≈6.82
    d. ≈7.41

20. How many nights has Mr. Jones slept within 1 standard deviation of the mean?
    a. 1
    b. 2
    c. 3
    d. 4

Answers to Odd-Numbered Multiple-Choice Questions

1. c
3. d
5. b
7. d
9. c
11. a
13. a
15. c
17. d
19. a

Part V Study Guide

The Normal Curve and Standard Scores

Part V Summary

Module 7

- The normal curve is a symmetric, bell-shaped curve that has an *inflection point* (meaning the curve bends) at 1 standard deviation above and below the mean. The shape of the normal curve is how a distribution with an infinite number of scores would appear. This means that the normal curve is theoretical as opposed to an actual distribution of scores. However, we can expect that scores will distribute themselves in a manner similar to that of the normal distribution, especially as the size of the sample grows beyond $n > 30$.
- The data sets of this book will rarely mimic the normal curve. However, the properties of the normal curve are *robust* to distributions that may violate its shape. This indicates that although our data set may not look perfectly normal, you can still use the special features of the normal curve to understand our sample.
- The benefit of using the normal curve is that you are able to determine the proportion (percentage) of scores that will fall within 1, 2, and 3 standard deviations of the mean. In a normal distribution, you can always expect that about 68% of the scores will fall within ($\pm$) 1 standard deviation of the mean; that about 95% of the scores will fall within ($\pm$) 2 standard deviations of the mean; and that about 99% of the scores will fall within ($\pm$) 3 standard deviations of the mean. This means that a score that falls 1 standard deviation above the mean will fall in the same place on the normal curve, regardless of what is being measured. Apart from knowing the proportion of scores at a specific standard deviation, you can use the normal curve to determine the exact proportion of the curve above and below any specific score.

Module 8

- *Standard scores* are scores that have been converted to a standard scale of measurement. This means that they can be compared with other scores that have been placed on the same scale. In other words, you can compare a person's height in feet with his or her weight in pounds by converting both values to standard scores. This is in contrast to raw scores, which are the scores on the original scale.
- The z *score* is a standardized score on the scale of standard deviation units. A z score of 1 means that the raw score fell 1 standard deviation above the mean. The sign of a z score (+/−) indicates the location of the score with regard to the mean

(− indicates below the mean, + indicates above the mean). The formula for finding a z score is as follows:

$$z = \frac{X - M}{s}$$

- The z-score formula is merely a conversion. First, you determine how far your score is from the mean in raw units $(X - M)$. Then, you determine how many standard deviations this distance is (dividing by s). This is similar to converting feet to inches or yards to feet.
- After converting a score to a z score, you are able to use it in conjunction with your knowledge of the normal curve because z scores are normally distributed. You can find the exact percentage of scores above your z score and the percentage of scores below your z score. These percentages are found using a normal curve table, which can be found in Appendix A of your textbook. The percentages on the normal curve table indicate the percentage of scores at and below a particular z-score value. To determine the percentage of scores above a z-score value, subtract the percentage from 1. Thus, a z score of 1.33 will be greater than or equal to 90.82% of the scores of the distribution, and less than 9.18% of the scores of the distribution. Finally, you can use the normal table to determine how many scores fall between a particular z score and the mean by subtracting .5 from the percentage found in the table.
- An important aspect of a z score is that it includes a distribution's central tendency (in the mean) and dispersion (standard deviation) in its calculation. This enables you to compare scores from completely different scales (such as tests in different classes) once they have been converted to z scores. Yes, this means you can finally compare apples with oranges!

Module 9

- You may need to adjust the scores in a distribution so that they are all higher, lower, more spread out, or more bunched up. This process is called a *transformation*. One type of transformation is to add a constant to every score in the distribution. Doing this would increase the mean by that constant as well. Alternatively, you may multiply every score by a constant. This would cause the mean to be multiplied by that constant as well.
- Transformations also affect measures of dispersion. Adding a constant to every score in the distribution would not affect the standard deviation. This is because this transformation doesn't move the scores closer to or farther from the mean. Rather, this process shifts the entire data set in a specified direction. However, if we were to transform our data by multiplying every score by a constant, we could expect the standard deviation to be multiplied by that constant. In multiplying the scores, we are moving each score further away from the mean and so we can expect the standard deviation to increase.
- It is important to note that regardless of the method of transformation you use on your data set, scores that are 1 standard deviation above the mean will always fall in the same place on a normal curve.
- Standardized scores are actually a transformation. You change any standardized score back to its raw score with the following formula:

$$\text{Raw} = s \times z + M$$

Learning Objectives

Module 7

- Know the history of the normal distribution
- Know the conditions under which data will be distributed normally
- Know the number of standard deviations in a normal distribution
- Know the percentage of cases falling between whole standard deviation values in a normal distribution

Module 8

- Know the advantages of standard scores over raw scores
- Understand the process of rescaling numerator units into denominator units
- Calculate a z score
- Use a normal curve table to determine percentages above, below, or between given z scores

Module 9

- Know the effect of score transformations on the percentage of cases falling above, below, or between various transformed scores
- Know the effect of score transformations on the mean and standard deviation of the set of scores
- Know the values of the mean and standard deviation for common standardized scores
- Convert scores from one type of standardized score to another

Computational Exercises

1. The mean grade on a French test was 74 with a standard deviation of 6. If you scored .5 standard deviations above the mean, what was your grade on the test?

2. What percentage of the normal distribution is greater than 2 standard deviations away from the mean? What percentage is greater than 1 standard deviation away from the mean?

3. What percentage of the normal distribution is greater than the mean? What percentage of the normal distribution is less than 2 standard deviations below the mean? What percentage of the normal distribution is within 1 standard deviation above the mean?

4. Using a distribution with a mean of 30 and a standard deviation of 4, find the z scores for the following scores:
   a. 28
   b. 14
   c. 39
   d. 4

5. Sana just took a very difficult cognitive psychology test and a very easy calculus test. On the psychology test, she earned a grade of 78 and found out the class had a mean of 72 with a standard deviation of 4. In contrast, she earned a grade of 87 on the calculus test and found out the class had a mean of 89 and a standard deviation of 6. How would you explain to Sana that she should feel good about her psychology test grade compared with her calculus test grade?

6. Using Question 5, the psychology teacher realizes that the test was very difficult and decides to give everyone in the class an additional 8 points. How does this affect Sana's grade, her standardized score, and your explanation about how she did on the test?

Here are the scores obtained by each of the 5 members of Tim's bowling team during last night's tournament. Use this information for Questions 7 to 9.

133
159
112
131
169

7. Convert all the scores to z scores. How many standard deviations above the mean was the person with the highest score?

8. What percentage of the normal curve is equal to or below the person with the lowest score? What about the person with the highest score?

9. The person who bowled a 159 thinks that he or she is better than the average bowler. What percentage above the mean is this person?

Two of the local Little League baseball teams want to have a contest to determine which team can catch more fly balls. Here are the means and standard deviations for number of fly balls caught by all the members of each team. Use this information for Questions 10 and 11.

Team A: Mean = 7.9; SD = 2.3

Team B: Mean = 8.3; SD = 3.9

10. Saul, Team A's best player, caught 10 fly balls. Ari, Team B's best player, also caught 10 fly balls. Which of these players is better, relative to the performance of their team?

11. Jose caught 4 fly balls and is on Team A. What proportion of the normal curve falls between Jose and the mean of Team A?

Here is the number of court cases a particular judge has seen per day for the past week. Use this information for Questions 12 to 15.

7
5
9
3
4

12. What is the z score for the day in which he saw the most cases? What percentage falls between this score and the mean?

13. The following week the judge is inundated with work and his caseload is doubled! What are the mean measures of central tendency and dispersion for the judge during this week?

14. The following week, the judge continues to have a busy time at work. However, it isn't as bad as the previous week (the week in which he had double the cases; Question 13) because he has 2 fewer cases per day. What are the measures of central tendency and dispersion for this week? (Using the original scores, multiply each score by 2 and then subtract 2 from each score.)

15. Using your information from Question 14, compare the busiest day (the day with the most cases) with the busiest day he had during the original week. Which day was more different from the average? How much more in terms of percentage of the normal curve?

Answers to Odd-Numbered Computational Exercises

1. 77

3. Fifty percent of the normal distribution is greater than the mean. Approximately 2.5% of the normal distribution is less than 2 standard deviations below the mean. Approximately 34% of the normal distribution is within 1 standard deviation above the mean.

5. You should tell her that she did much better on the psychology test relative to everyone else as she had a z score of 1.5. However, on the calculus test, she didn't do as well compared with everyone else as she had a z score of –0.33.

7. $M = 140.8$, $s = 20.56$

X	Z
133.00	−0.38
159.00	0.89
112.00	−1.40
131.00	−0.48
169.00	1.37

The person with the highest score is 1.37 standard deviations above the mean.

9. Thirty-one percent of the normal distribution falls between a z score of 0.89 and the mean of 0.

11. Approximately 46% of the normal curve falls between Jose and the mean of Team A.

13. The mean for this week would be 11.2 and the standard deviation would be 4.3. Each of these scores is multiplied by 2.

15. The z score for the busiest day with the original scores was 1.58. The z score for the busiest day from Question 14 was also 1.58. Although the scores changed, their position in the distribution remains the same.

True/False Questions

1. The majority of the scores of the normal curve occur within 1 standard deviation of the mean.

2. The end points of the normal curve represent scores with a frequency of 0.

3. You can still use the assumptions associated with the normal distribution with distributions whose shape slightly deviates from normality.

4. Approximately 50% of the scores occur within 3 standard deviations of the mean.

5. Standardized scores tell you the position of a score in reference to all the other scores in a distribution.

6. A *z* score of 1.5 corresponds to a score that is 1.5 standard deviations above the mean.

7. A *z* score only tells you the location of a score in relation to the mean.

8. A *z* score is in the scale of variance units.

9. Converting all the scores in a distribution to *z* scores automatically converts the distribution to a normal distribution.

10. The average golf score for members of a golf team was 4. It was later found out that the team was cheating on their scores and all the members were reducing their scores by 3 strokes. This indicates that the true mean of the golf team was 7.

11. The proportion of the normal curve below a *z* score of 0.12 is .45.

12. You can compare scores from entirely different groups after converting the scores to *z* scores.

13. Brand A sells an average of 5.4 units per day with a standard deviation of 1.2. Brand B sells an average of 3.2 units per day with a standard deviation of 2.5. A *z* score of 1 would represent different raw scores but indicate a score that fell in the same place on the normal curve.

14. Multiplying all the scores in a distribution by a constant will cause the mean to be multiplied by that constant but not affect the measure of dispersion.

15. The mean of IQ scores is 100 with a standard deviation of 15. The mean of MMPI scores is 50 with a standard deviation of 10. An IQ score of 85 and an MMPI score of 40 fall in the same place on the normal curve.

Answers to Odd-Numbered True/False Questions

1. True
3. True
5. True
7. False
9. False
11. False
13. True
15. True

Short-Answer Questions

1. What is an inflection point? Why is it important in the normal curve?
2. What aspect of the normal curve makes it very useful in statistics?

3. With regard to statistics, what does it mean for something to be robust?

4. What does it mean for a score to fall at the 85th percentile of the normal curve?

5. What two pieces of information can be found from a z score?

6. You want to compare the average amount of nail polish used by a nail salon in a week with the average amount of lettuce sold per week by a grocery store. What type of scores could you use to make this comparison? What is it about this type of score that helps you make this comparison?

7. How does the formula for a z score "rescale" raw scores?

8. If you were to find that you have a score that is worse than 2.5% of the population, where would you fall in the normal distribution?

9. What will the mean and standard deviation of a distribution that has been converted to z scores always be?

10. How are the mean and standard deviation affected by adding a constant to all the scores in a data set?

11. How are the mean and standard deviation affected by multiplying all the scores in a data set by a constant?

12. What does adding a constant to each score in a data set do to z scores? Why is this?

13. What does multiplying each score by a constant do to z scores? Why is this?

14. You design a measure to assess your classmates' opinion of this course that ranged from −10 to 10. However, the teacher asks to see the scores and you don't want to hurt his or her feelings by showing that some of the students provided negative ratings. How could you fix this problem while not actually changing the location of the scores in the normal distribution?

15. The students in Course A earn a mean grade of 85 with a standard deviation of 2.32 on the first test. The students in Course B earn a mean grade of 93 with a standard deviation of 3.84. You are enrolled in both courses and discover that your z score for Courses A and B was 0.48. Does this mean that you have the same grade in both courses? Why or why not?

Answers to Odd-Numbered Short-Answer Questions

1. An inflection point is the place on the normal curve where the frequency curve drastically changes direction. On the normal curve, the first major inflection point represents 1 standard deviation above and below the mean.

3. It means that you can still use the statistical procedure although some of the required assumptions of the procedure have been violated. For example, you can still use the rules of normal curve distribution with data that do not perfectly conform to the shape of the normal curve.

5. The distance of the score from the mean in standard deviation units and the direction of the score from the mean (either above or below).

7. A z score tells you the location of a score in relation to the mean in standard deviation units. This is obtained by first subtracting the score from the mean (numerator), resulting in a deviation score. Then, this deviation score is converted to standard deviation units by dividing by the standard deviation.

9. The mean will always be 0 and the standard deviation will always be 1.

11. The mean and standard deviation are both multiplied by that constant as well.

13. This would increase all the z scores. This is because multiplying by a constant "expands" the distribution, which means that each score will move further from the mean.

15. The grades are not the same in both courses because the means and standard deviations of the distributions are different. However, it does mean that your score fell in the same location on the normal curve in both classes.

Multiple-Choice Questions

1. Over 99% of the scores within a distribution fall within how many standard deviations of the mean?
   a. 0
   b. 1
   c. 2
   d. 3

2. As you move further away from the mean in a normal distribution, the frequency of scores
   a. Increases
   b. Decreases
   c. Stays the same
   d. The answer depends on the data set that is used

3. What proportion of the normal curve is greater than a z of 0.34?
   a. .3669
   b. .4521
   c. .6331
   d. .7892

4. What proportion of the normal curve falls between a z score of −1.2 and the mean?
   a. .0214
   b. .3849
   c. .7842
   d. .6254

5. What proportion of the normal curve falls between $z = 0.64$ and $z = 0.89$?
   a. .2389
   b. .3133
   c. .0744
   d. .9256

6. What proportion of the normal curve falls between $z = -0.93$ and $z = 1.6$?
   a. .8238
   b. .3238
   c. .7690
   d. .4452

7. What proportion of the normal curve falls outside $z = 1.5$ and $z = 0.34$?
   a. .6999
   b. .6331

c. .9331
d. .3001

8. If a distribution of scores has a mean of 15 and a standard deviation of 1.5, what is the z score for a score of 12?
   a. −0.5
   b. −1
   c. −1.5
   d. −2

9. The director of a new horror movie wants to know how scared her daughter was while watching the film. She asks her daughter to rate her fear on a scale of 1 to 15. If the average rating of fear for this movie is $M = 6$ with an $SD = 0.75$, where in the normal distribution would the director's daughter be if she gave a rating of 5?
   a. Higher than 25.23%
   b. Lower than 78.65%
   c. Higher than 9.18%
   d. Lower than 36.47%

10. The local police department wants to determine the amount of respect that the local citizens have toward them. They obtain data from all the residents and find that the $M = 3$ and the $SD = 2.7$. In looking through the data, they notice that one participant rated his or her respect level at 9. What is the z score for this person?
    a. 1.22
    b. 2.22
    c. 0.79
    d. −1.62

11. The raw scores of a distribution have an $M = 78$ and an $SD = 12.3$. You obtain a z score of 2.3. What is your corresponding raw score?
    a. 100.54
    b. 94.24
    c. 106.29
    d. 145.63

12. In a recent race, Devon is bragging that he was faster than anyone else. In fact, he goes so far as to say that he finished 3.1 standard deviations ahead of everyone else. You are skeptical of how big a difference this really is and find that the time it took for everyone to finish the race was $M = 93$ with an $SD = 2$. What was Devon's time to finish the race?
    a. 47.5
    b. 110
    c. 94.6
    d. 99.2

13. A car advertisement states that the manufacturer's new models are able to accelerate at a rate that is 2 standard deviations above the mean. If the average acceleration of a car (in miles per hour) is 50 with an $SD = 5.3$, what is the acceleration of the car being advertised?
    a. 45.2
    b. 55.3
    c. 60.6
    d. 65.7

14. A makeup company is attempting to lower its prices in order to improve sales. They decide to charge $5 for all their products. If eyeliner has $M = \$6.50$; $SD = \$0.35$ and foundation has $M = \$7.45$; $SD = \$0.78$, which of these products has the bigger relative price reduction?
    a. Eyeliner
    b. Foundation
    c. Both have an equal reduction
    d. Need more information

15. Two brothers are having an eating competition. Mike is seeing how many hot dogs he can eat in a minute and Jeff is seeing how many hamburgers he can eat in a minute. If Mike eats 7 hot dogs and usually eats $M = 4.5$; $SD = 1$, and Jeff eats 6 hamburgers and usually eats $M = 3.2$; $SD = 2.1$, who ate more relative to his normal amount of consumption?
    a. Mike
    b. Jeff
    c. Both ate the same amount
    d. Can't determine

16. Three electronics companies are trying to beat each other's price on car stereos. Here is a table with the average original price, standard deviations, and the current sale price. Who is offering the best deal?

Company	Mean Original Price ($)	SD ($)	Mean Sale Price ($)
A	450	23	389
B	375	12	359
C	475	57	401

    a. Company A
    b. Company B
    c. Company C
    d. All the companies are offering comparable price drops

17. After such a sale, Company B needs to increase their profits. As a result, they increase the original price on all their models by $10. What is the new mean price of their stereos in dollars?
    a. 375
    b. 385
    c. 365
    d. Need more information

18. A group of students are misbehaving and so a teacher states that they will all lose 5 points on their next test. If the students all earned an average grade of 72 with a standard deviation of 4, what would the new mean and standard deviation be with the penalty?
    a. $M = 72$; $SD = 9$
    b. $M = 77$; $SD = 4$
    c. $M = 77$; $SD = 9$
    d. $M = 72$; $SD = 4$

19. You are playing a number game with friends. You give them a list of 10 numbers with an $M = 62$; $SD = 4$. You then tell them to add 4 to every number and then multiply

every number by 2. They are amazed when you immediately tell them the new mean and standard deviation after making these changes. What are these new values?
   a. $M = 66$; $SD = 8$
   b. $M = 102$; $SD = 12$
   c. $M = 132$; $SD = 8$
   d. $M = 132$; $SD = 16$

20. Your friends think you cheated at the number game you played in the previous question. To prove them wrong, you tell them you are willing to do it again and will even let them provide you with 20 random numbers. The descriptive statistics of these new numbers are $M = 70$ and $SD = 15$. To make it extra hard, your friends tell you that you should first divide all the numbers by 5, then multiply them all by 3, and finally subtract 6 from all scores. You astonish your friends by correctly telling them the mean and standard deviation would be
    a. $M = 42$; $SD = 3$
    b. $M = 36$; $SD = 9$
    c. $M = 90$; $SD = 7$
    d. $M = 14$; $SD = 5$

Answers to Odd-Numbered Multiple-Choice Questions

1. d
3. a
5. c
7. a
9. c
11. c
13. c
15. a
17. b
19. c

Part VI Study Guide

Probability

Part VI Summary

Module 10

- Probability theory helps us understand the properties of the normal curve. As mentioned in previous modules, the properties of the normal curve allow us to perform most inferential statistics. Therefore, it is important to have an understanding of probability. However, probability is a vast topic, and this book will only touch on a few key elements that are relevant to the statistics in this textbook.
- *Probability* is the chance that you will obtain a specific outcome (e.g., Outcome A) out of many different outcomes (e.g., Outcomes A, B, C, or D). Probability is frequently expressed as a proportion, representing the chances of obtaining the specified outcome divided by the total number of outcomes. In the previous example, the probability of obtaining Outcome A, expressed as $p(A)$, would be 1/4. This is because there is one Outcome A, but four possible outcomes.
- The previous example assumes that the chance of getting any particular outcome (A, B, C, or D) is equal. This is referred to as an *equally likely model*. It is important to note that not all situations conform to the equally likely model. For example, if there were two Outcome As (A, A, B, C, D), then the chances of getting an A would be higher than the chances of getting another letter. However, the majority of statistics for this text are based on the equally likely model.
- *Mutually exclusive outcomes* indicate that you can only obtain one outcome per trial. If the possible outcomes are A, B, C, and D and you obtain Outcome B, then you cannot have an outcome of A, C, or D on that trial. Although it is possible for outcomes to not be mutually exclusive, the majority of the statistics in this text will deal with mutually exclusive outcomes.
- The *addition theorem* states that the probability of any of the possible outcomes occuring on a *single trial* is equal to the sum of their individual probabilities. Returning to our example of A, B, C, D, the probability of obtaining either an A or B on a single trial is equal to the probability of obtaining A plus the probability of obtaining B. Therefore, this probability would be 1/4 + 1/4 = 1/2.
- Another consideration in probability is how the outcome of one trial affects the outcome of another trial. *Independent outcomes* are those in which the outcome of one trial has no impact on the outcome of another trial. If you were to flip a coin twice, you would have a 50% chance of obtaining a head on the second flip, regardless of what side the coin landed on in the first flip. The independence of outcomes on successive trials will play a role in which type of statistical test you will use. This will be covered at length in later chapters.

- Finally, the *multiplication theorem* states that the probability of obtaining specified outcomes in a *series of trials* is the product of the individual probabilities of each specified outcome. Using the A, B, C, D example, this means that if the probability of obtaining A on the first trial is 1/4 and the probability of obtaining B on the second trial is 1/4, then the probability of obtaining A and then B would be 1/4 × 1/4, or 1/16.
- It is important to note that when you have only a limited number of trials, you cannot guarantee any certain outcome. In other words, although we have 4 outcomes, each having 1/4 chance of occurring, we should not expect each outcome to occur once and only once in four different trials. This means that probabilities are *theoretical*, meaning probabilities are what you expect to happen (you expect four different outcomes). Actual results are *empirical*, meaning they actually happen (you may get two As, one B, and one D). However, as you increase the number of trials, your empirical findings will more closely resemble your theoretical expectations. If we repeated the four-outcome example 200 times, you could expect close to 50 As, 50 Bs, 50 Cs, and 50 Ds (although not exactly these numbers).
- Research in the social sciences frequently focuses on comparing two different groups, such as a treatment and a control (no-treatment) group. Using the equally likely model, you would expect there to be no difference between the groups, that the scores of the treatment group = scores of control group. However, if the treatment is effective, in which case the outcomes are not expected to be equally likely, then you would notice a difference in scores between these groups. This difference would mean that the probability of falling in one specific group is not equally likely; it depends on the person's score. This suggests that the treatment is effective.

Module 11

- *Dichotomous outcomes* are outcomes that have only two possible outcomes. The results from dichotomous outcomes with a probability of .5 per outcome tend to be distributed in a pattern that resembles the normal curve. Recall that p represents the probability an outcome will occur and q represents the probability an outcome will not occur.
- This pattern is referred to as the binominal distribution. By using this pattern, it is possible to determine the probability that you would obtain any possible combination of outcomes across any number of trials. There are two methods to determine these probabilities. The first requires you to add all the probabilities of obtaining a combination with your desired number of outcomes. For example, if you wanted to determine the probability of obtaining at least 2 heads in three coin flips, you would add the probability of all ways to earn exactly 2 heads and then all the ways of earning 3 heads. This sum would provide you with the probability of earning at least 2 heads in three flips.
- However, the previously mentioned process is rather tedious. An alternative method is using the binomial formula, which is $(p + q)^N$. When using this formula to calculate probabilities, it is necessary to expand the formula through the use of a process called *binomial expansion*. This process expands the formula to appear as follows:

$$(p+q)^N = p^N + Np^{N-1}q + \frac{N(N-1)}{1 \times 2}p^{N-2}q^2 + \cdots + q^N$$

This formula provides the probability of obtaining all combinations of outcomes in N trials. If you were to graph the results of this formula, it would bear a strong resemblance to the normal curve.

- Apart from binomial expansion, the text provides a table (Appendix B) that will provide you with the probabilities for obtaining each specific outcome for a number of trials (up to 8). Also, the table provides information for a number of different probabilities for p and q.
- The binomial distribution is relevant to the social sciences, and specifically experimentation, because it provides a set of expected probabilities from an unlimited number of trials. This means that you do not need to conduct an experiment hundreds of times to feel confident about the results. Rather, you can consult the normal curve to determine the probability of obtaining such a result. For example, you are interested in determining if a group of people are generous. You ask 10 people to donate to a charity, with the expected response of either a yes or no. Also, you expect the probability of a person responding with a yes to be .5. It would be highly unlikely that you would obtain 10 yes responses. However, if you did, you may conclude that the chance of obtaining a yes response from this group is higher than expected. You could further conclude that the reason you got 10 yes responses is because this group of people is generous.

Learning Objectives

Module 10

- Distinguish between mutually exclusive, embedded, and overlapping outcomes on a single trial
- Distinguish between independent and dependent outcomes in a series of trials
- Know the conditions under which the addition theorem applies
- Distinguish between theoretical and empirical outcomes
- Understand the use of empirical versus theoretical data in making inferences

Module 11

- Calculate probability by listing all possible outcomes
- Calculate probability by expanding the binomial formula
- Find probability by using a binomial table
- Understand the relationship between likelihood in a dichotomous model and the shape of the outcome distribution

Computational Exercises

1. Elba was asked by her partner to pick up some tomato sauce at the grocery store. Unfortunately, she forgot what brand she was supposed to purchase. The grocery store has 10 different types of tomato sauce and Elba doesn't remember anything about the type she was supposed to buy (her chance of picking any of the 10 is equally likely). What is the probability that Elba will pick the correct brand? If Elba were to buy 2 brands, what is the probability that at least one of them will be the correct brand?

2. Samantha wants to rent a comedy from the movie store. In browsing the comedy section, she sees four movies that she would really like to see, but is uncertain of which one to pick. She calls her roommate and asks for her roommate's opinion on

the matter. If Samantha were to pick a title but not tell her roommate (so that the selection they would both make would be mutually exclusive), what are the chances that they would pick the same title?

3. You are attending a conference in which there is a representative from one of each of the 50 United States (you are not a representative). You are introduced to one of the representatives. What is the probability that
   a. This person is from one of the 50 United States?
   b. This person is from New York?
   c. This person is from California, Texas, or Rhode Island?

4. You are playing a dice game with your friend that uses 3 six-sided dice.
   a. What is the probability that you will roll a 1 with the first die, a 2 with the second, and a 3 with the third?
   b. What is the probability that you will roll a 6 with the first die, a 6 or a 4 with the second die, and a 1 with the third die?
   c. What is the probability that you will roll a value greater than 3 on the first die, a value of 3 or less on the second, and a 4 on the third?

5. You and a few of your friends are preparing for the beach and discussing which color beach towel each person should bring. You have a total of 15 beach towels: 3 are blue, 2 are green, 6 are red, and 4 are pink.
   a. If you close your eyes and grab a towel, what is the probability that it will be red or pink?
   b. If you took a green towel, what is the probability that your friend (who also takes a towel without looking) will take the other green towel?

6. At the beach, you encounter a person playing a game. He has two cups with an almond under one of them. He shuffles the cups and asks you to pick which one has the almond. However, he shuffles the cups so quickly that you lose track of the almond and now have to guess at random. List all the possible combinations of selecting the almond across 3 trials.

7. What is the probability for all the combinations listed in Question 6? What is the *p* and *q* for each combination?

8. Sketch a graph of the probabilities for getting a hit across the 3 trials in Question 6. If the person with the cups offers you a prize for guessing it three times in a row, but states that you must pay him if you miss all three times in a row, is this a fair deal?

9. You recently ordered 4 replacement parts for a lamp you purchased. The sales representative tells you that each part will come either next Wednesday or next Thursday, and all parts have an equal chance of coming on either day. Using binomial expansion, what are the probabilities of getting your parts sooner rather than later? (Wednesday is a hit, Thursday is a miss.)

10. You discover that you are missing an additional 2 pieces for the lamp in Question 9. Fortunately, the sales representative is able to ship these pieces with the rest and you are slated to receive them also on next Wednesday or Thursday. Using binomial expansion, what are the probabilities of getting your parts (6 in total) sooner rather than later?

11. Surprisingly, you do not receive any of your parts on either day (from the previous two questions). Upset, you call the company and ask them when you should expect them. They say that you should expect all 6 parts on either Monday or Tuesday.

However, the company has mailed with an express delivery and there is a greater chance (.75) you will obtain them on Monday. What is the probability of obtaining all 6 parts on Monday using the table in Appendix B?

12. Complete a binomial expansion for 7 trials.

13. David is practicing his dart throwing on a dart board whose area is 10% bull's-eye and 90% miss. If he isn't very good and is just doing his best to hit the dart board (which he always will), what is the probability he will hit the bull's-eye at least twice in 7 throws? (Use the table in Appendix B.)

14. Using the example from above, David has improved his accuracy and now has a 20% chance of hitting the bull's-eye. If he throws 5 times, what is the probability that he will hit the bull's-eye once, twice, or three times?

15. David feels so confident in his dart-throwing ability (now 30% of hitting the bull's-eye) that he decides to impress his friend Shakra. He tells Shakra that she must pay him a dollar if he can hit the bull's-eye at least twice in eight throws. Should Shakra take this bet? Use the table in Appendix B to help you make the decision.

Answers to Odd-Numbered Computational Exercises

1. The probability she will pick the correct brand is 1/10 or .1. The probability she will pick one of the correct brands if she bought 2 is 2/10 or .2.

3.
   a. 50/50
   b. 1/50
   c. 3/50

5.
   a. $\left(\dfrac{6}{15}\right)\left(\dfrac{4}{15}\right) = \dfrac{24}{225}$

   b. $\left(\dfrac{1}{15}\right)\left(\dfrac{1}{14}\right) = \dfrac{1}{210}$

7.

Combination	p	q
MMM	.125	.875
HMM	.125	.875
MHM	.125	.875
HHM	.125	.875
MMH	.125	.875
HHH	.125	.875
MHH	.125	.875
HMH	.125	.875

$p(2 \text{ hits}) = 4/8 = .5$; $p(1 \text{ hit}) = 1/8 = .125$

9. $.5^4 + 4(.5^3)(.5) + \frac{(4)(3)}{(1)(2)}(.5^2)(.5^2) + \frac{(4)(3)(2)}{(1)(2)(3)}(.5^1)(.5^3)$
$+ \frac{(4)(3)(2)(1)}{(1)(2)(3)(4)}(.5^4) = .0625$

11. $N = 6$, $p = .75$, $q = .25$, $p(6) = .1780$

13. $N = 7$, $p = .10$, $q = .9$, $p(\geq 2) = .1498$

15. $N = 8$, $p = .30$, $q = .7$, $p(\geq 2) = .7447$. She should not take the bet.

True/False Questions

1. The probability of all possible outcomes must sum to 1.
2. Mutually exclusive outcomes indicate only one outcome is possible per trial.
3. Two soccer teams are playing each other. One of the teams has much better players than the other and is favored to win. The probability of this team winning is the same as the probability that the team with the weaker players will win.
4. The probability that one outcome will occur or another outcome will occur is the product of these two individual probabilities.
5. If the probability that you will pass a math test is .90 and the probability you will pass a history test is .8, the probability that you will pass both tests is .72.
6. The addition theorem applies to outcomes over many trials.
7. You and a friend are drawing straws from a bag that contains 6 straws. You draw 1 straw and keep it. Then, your friend draws 1 straw and keeps it. Then you draw again and so on until there are no straws left. These trials each have independent outcomes.
8. Probabilities are empirical, whereas observed results are theoretical.
9. Theoretical principles are used as standards by which empirical results are compared.
10. Dichotomous outcomes always have two outcomes.
11. If the probability of two outcomes is equally likely, then binomial distribution resembles the normal distribution after many trials.
12. If two outcomes are not equally likely, then the binomial distribution will appear skewed.
13. Statistics and probability will provide you with a clear understanding of why an outcome occurred.
14. You and a friend are playing catch. If your friend has a .4 chance of missing a catch, then the probability of him or her missing the next two throws is .8.
15. The majority of events in life are dichotomous with equally likely outcomes.

Answers to Odd-Numbered True/False Questions

1. True
3. False

5. True

7. False

9. True

11. True

13. False

15. False

Short-Answer Questions

1. What does a probability of .75 indicate?

2. What does it mean for outcomes to be equally likely? Can there be more than 2 equally likely outcomes?

3. What are mutually exclusive outcomes? Are mutually exclusive outcomes more common or rare when using statistics for the social sciences?

4. Compare and contrast when the addition theorem and the multiplication theorem should be used.

5. Sam is learning to play a song on the saxophone. Each day he practices for 2 hr. Is the probability that he will correctly play the song each day an independent outcome? Why or why not?

6. If you were conducting an experiment in which you wanted to obtain the public's opinion on whether they enjoyed a recent movie, why would it be important to have independent outcomes? (*Hint*: Consider each person you would ask a separate trial.)

7. Would you consider your performance on tests in this class as independent outcomes? What about your performance on tests in this class as compared with tests in other classes?

8. What does it mean for something to be empirical as opposed to theoretical?

9. What does the expression "Nine times out of 10 . . ." really indicate?

10. What are the three criteria that must be met in order for a set of outcomes to be considered dichotomous?

11. Using the binominal distribution, how does the probability change as you increase the number of "hits" from 0, if the outcomes are equally likely?

12. Why is the binominal distribution important to other inferential statistics?

13. When would a set of dichotomous outcomes produce a skewed binomial distribution?

14. Are the following dichotomous situations equally likely or not?
    a. Winning a game at a carnival
    b. Passing a test
    c. Enjoying a movie
    d. Flipping a coin
    e. Dropping a piece of toast butter side up or dry side up.

15. You are comparing the effect of a new therapy for autism. At the start of the study, you state that if the treatment doesn't work, then those who receive the therapy will be no better than those who did not receive the therapy (the theoretical difference in symptoms between the two groups based on probability would be 0). However, you discover that those who had the treatment have fewer symptoms than those who did not. Furthermore, the difference in symptoms between the groups is quite large and the probability of observing this difference is .03. Can you conclude that your treatment works? Why or why not?

Answers to Odd-Numbered Short-Answer Questions

1. It indicates that theoretically, one set of outcomes occurs 75% of the time and another 25% of the time.

3. Mutually exclusive outcomes are those in which only one outcome is possible per trial. These are more common when using statistics from the social science perspective.

5. These trials are not independent. Each day that Sam practices, his ability to play the song improves. Thus, the probability that he will correctly play the song today depends on how well he played the song the previous day (assuming he is improving in his playing).

7. You would not consider your performance on tests in this class as independent outcomes as your ability to do well in statistics might improve with each successive trial, or it might get worse due to dependence on previously ill-understood material. Your performance across classes could be considered independent outcomes as your knowledge of statistics may not influence your knowledge of history. However, an argument for them being not independent could be made as study skills you develop for statistics may carry over to other classes.

9. This would suggest that the probability of obtaining a specific result is .9. In other words, it indicates that the chance of obtaining one specific outcome is rather high.

11. The probability increases as you increase the number of hits from 0. However, after obtaining approximately an equal number of hits and misses, the probability of obtaining more hits than misses decreases. Eventually, the probability of obtaining all hits is equal to the probability of obtaining no hits.

13. A skewed distribution is produced when the probabilities for a dichotomous outcome are not equally likely.

15. You can feel confident that your treatment had an effect on symptoms, but you cannot be certain that it worked. The fact that the chance of obtaining such a difference is very low does provide strong evidence that the treatment worked. However, it does not prove that it was effective as you have not tested all the people with autism with your treatment.

Multiple-Choice Questions

1. If the probability of an event occurring is .45, what is the probability of the event not occurring?
   a. .45
   b. .35

c. .50
d. .55

2. If the probability of an event occurring is .1 and alterative outcomes are equally likely, how many possible outcomes are there?
   a. 1
   b. 2
   c. 5
   d. 10

3. What is the probability that you will select the correct choice for this question without knowing what you are being asked?
   a. .5
   b. .1
   c. .25
   d. 1

4. Which of the following is false: Dichotomous outcomes always
   a. Have two solutions
   b. Are equally likely
   c. Are mutually exclusive
   d. Are independent

5. There are 20 questions in this section. If you were to select a choice completely at random for all 20 questions, what is the probability that you would get all the answers correct?
   a. .25
   b. .1
   c. .01
   d. <.001

6. If you were to draw a card from a deck of 52 cards, what is the probability that it would be a king or a heart?
   a. 1/52
   b. .02
   c. >.001
   d. These outcomes are not mutually exclusive

7. If you were to draw a card from a deck of 52 cards, what is the probability that it would be a face card (jack, queen, king)?
   a. .23
   b. .02
   c. .06
   d. .01

8. The multiplication theorem cannot be applied under which circumstances?
   a. Single trials
   b. Mutually exclusive outcomes
   c. Independent outcomes
   d. Dichotomous outcomes

9. Anita is a clumsy waitress and has a tendency to drop things. If the probability that a cup will fall on its side is .5, right side up is .25, and upside down is .25, what is the probability that the 3 cups Anita just dropped will land right side up?
   a. .75
   b. .50

c. .02
d. .00

10. You are starring in a play that depicts a 1950s gangster who habitually flips a coin. You begin to count the number of times you obtain tails in each series of flips. If your recent practice session included 4 flips, how many different ways would there be to obtain 2 or more tails?
    a. 9
    b. 10
    c. 11
    d. 12

11. Using the above example (Question 10), what is the probability that you would obtain 3 or more tails in 4 flips?
    a. .2500
    b. .3125
    c. .0625
    d. .3750

12. On a multiple-choice question, there are 5 options. If you are able to eliminate two of the possible choices, what is the probability of selecting the correct answer from the remaining choices?
    a. .20
    b. .33
    c. .50
    d. .75

13. The weatherman forecasts that the chance it will rain over the next week is .4 per day. What is the probability that it will rain for exactly 3 days?
    a. .2903
    b. .7009
    c. .2160
    d. .6000

14. Using the probabilities in Question 13, what is the probability that it will rain for at least 4 days?
    a. .1935
    b. .2881
    c. .2897
    d. .2679

15. Using the probabilities in Question 13, what is the probability that it will rain for 1 or 3 days?
    a. .2903
    b. .1306
    c. .2548
    d. .4209

16. A salesman states that his cough medicine is guaranteed to work! He reports that he gave the medicine to 7 people and they all immediately lost their cough. You are skeptical of this salesman and want to tell him the probability that this would happen. What is the probability of this happening assuming that the medicine had an equal chance of curing the cough or not curing the cough?
    a. .0547
    b. .0078

    c. .1641
    d. .0098

17. The salesman in Question 16 confesses that it didn't cure all the coughs. However, he states that there is a 95% chance it would cure your cough. If you were to take this medicine 5 times and this probability is true, what is the probability that the medicine would cure your cough every time?
    a. .9514
    b. 1
    c. .5054
    d. .7738

18. As it turns out, the salesman from Question 16 was lying about his cough medicine. In fact, it only helped relieve coughing 10% of the time. If you were to give this to 3 people, what is the probability that it would reduce at least one person's cough?
    a. .7290
    b. .2710
    c. .2430
    d. .0280

19. Dichotomous outcomes are usually going to be on which scale of measurement?
    a. Nominal
    b. Ordinal
    c. Interval
    d. Ratio

20. How would a binominal distribution in which the probability of a hit is .9 appear after an infinite number of trials?
    a. Normally distributed
    b. Positively skewed
    c. Negatively skewed
    d. Bimodal

Answers to Odd-Numbered Multiple-Choice Questions

1. d
3. c
5. d
7. a
9. c
11. b
13. a
15. d
17. d
19. a

Part VII Study Guide

Inferential Theory

Part VII Summary

Module 12

- *Descriptive* statistics are used to describe the participants in a sample. Descriptive statistics include means and standard deviations. *Inferential* statistics are used to learn (or infer) about a larger *population* through the use of a sample.
- To use inferential statistics, your sample must be *representative*, meaning it has similar properties to that of the population. A representative sample should be similar in almost all respects to the population of interest.
- Representative samples are drawn through different *sampling* techniques. One of the most commonly used sampling techniques is simple random sampling. *Simple random sampling* is the process in which every person in the population has the opportunity to be included in the sample and those who are included are chosen at random. Another commonly used sampling method is stratified random sampling. *Stratified random sampling* involves dividing the population into categories and then drawing random samples from those categories. In this process, the proportion of each category in the population is usually maintained in the sample. For example, if you knew that the population of your school was 60% male and 40% female, you would want your sample to contain 60% males and 40% females. *Cluster sampling* involves selecting preexisting groups of people, such as classes or teams. Finally, *convenience sampling* is taking a sample from those subjects who are available for the study at that particular time. Convenience sampling may not provide you with a representative sample.
- In research, there are typically two main variables of interest. The *independent variable* (IV) is the variable that is manipulated by the researcher and expected to bring about a difference in another variable. The *dependent variable* (DV) is the variable that is believed to be changed by the IV. The DV is what is observed in a research study.
- Although the IV and DV are the primary variables of interest, there are additional variables that can pose problems to research designs. *Extraneous or confounding variables* are those that are associated with the IV and have an unwanted effect on the DV. This may cause the researcher to incorrectly conclude that the IV caused a change in the DV.
- A *hypothesis* is an educated guess about what is expected to happen in a research study. Generally, these statements are worded in terms of the effect the IV will have on the DV. *Research or alternative* hypotheses indicate that the IV will have an effect on the DV. These hypotheses can take two forms, directional or nondirectional. *Directional* hypotheses indicate that the IV is thought to affect the DV in a specific

direction (e.g., make the DV increase or make the DV decrease). *Nondirectional* hypotheses do not indicate a specific direction in which the IV may affect the DV. Rather, they state that the IV will cause the DV to change, meaning that the DV may increase or decrease.
- The *null hypothesis* states that the IV will have no effect on the DV. In other words, the null hypothesis states that there will be no difference in the DV, regardless of what is done with the IV.
- The null and alternative hypotheses provide a dichotomous situation for all research questions. Either the IV does not have an effect on the DV (the null hypothesis) or the IV does have an effect on the DV (the alternative). Research can never prove the alternative to be true because research will never be able to accurately assess an entire population. Thus, you can only say that you have not supported the null hypothesis. In doing so, you have supported the alternative hypothesis.

Module 13

- In a perfect situation, sample statistics would mirror population parameters. However, you can expect some difference between your sample statistics and your population parameters when conducting research. The extent that these two measurements differ is referred to as *sampling error*. Sampling error is the deviation between a sample statistic and a population parameter that is attributed to chance as opposed to other variables. Since this difference is attributed to chance, it is expected that this difference between a sample statistic and a population parameter will be small.
- In research, you manipulate your IV and expect to observe a change in the DV. Under the null hypothesis, we expect that there will be no change in the DV. In other words, the null hypothesis is supported when the deviation between a population parameter and a sample statistic is minimal as this difference is thought to be attributed to sampling error. However, when a sample statistic differs greatly from the population parameter, we reject the null hypothesis because we feel that the difference is caused by the manipulation of the IV. A large difference, one that is not expected by mere chance, is called a *significant difference*. However, this does not prove that the IV caused the difference. It is still possible that the large difference between the sample and population is due to chance. This is the difficulty when working with probabilities.
- When you make a decision about whether to reject or retain (support) the null hypothesis, you may be making a correct decision (hit) or an incorrect decision (miss). As mentioned before, large differences between the groups are thought to be related to the IV. If we obtain a large difference between our groups, and the difference was actually caused by the IV, then we have made a correct decision to reject the null hypothesis (hit). However, we may also obtain a large difference due to mere chance. This would occur if our sample happened to be unrepresentative of the original population. Unfortunately, unaware that the difference between our groups is due to chance, we would still reject the null hypothesis. This situation is referred to as a *Type I error,* where we have incorrectly rejected the null hypothesis. Type 1 errors are also symbolized as *alpha* or α.
- Just as you may mistakenly reject the null hypothesis in a Type 1 error, you may also mistakenly retain the null hypothesis. It may be the case that you have selected a sample that is highly uncharacteristic of the treated population. This sample may be more representative of the untreated population and so the difference between the groups would be rather small. Remember, a small group difference is thought to be attributed to sampling error and so you would incorrectly retain the null hypothesis. When the null hypothesis is retained incorrectly, you have committed a *Type 2 error*. Type 2 errors are symbolized as *beta* or β.

Module 14

- It is important to remember that there is always the chance of committing a Type 1 or a Type 2 error in statistics and that this threat will be prevalent throughout this text and all hypothesis testing. Thus, the best way to feel confident about your results is to replicate your study with many different samples.

Module 14

- The formula for z scores is actually the foundation for all types of inferential statistics. The numerator of this formula $(X - M)$ is comparing what you observed (X) to what you expected (M). The mean is the expected value because it is the most frequently occurring score when the data are normally distributed. The denominator (s) is the expected deviation of a score from the mean, when the null hypothesis is true. In other words, the standard deviation is how much random error is expected.
- Using this interpretation of the z-score formula, a z score could be expressed as a hypothesis test. The null hypothesis would state that you expect no difference (or minimal difference) between your observed score (X) and your expected score (M). You then determine how much of a difference you observed as compared with how much was expected (s). If this comparison tells us that we have a large difference, then we have grounds to reject the null hypothesis. If this comparison tells us that we have a small difference, then we would retain the null hypothesis. We can use probabilities to help us determine whether or not we have a big difference or a small difference. Using Appendix A, we can determine the probability that we would obtain this specific score. For example, an obtained z score of 1.22 indicates that you would obtain a score higher than this 11.12% of the time and a score less than this 88.88% of the time. Based on these probabilities, you can decide whether or not the difference is large enough to reject the null hypothesis.

Learning Objectives

Module 12

- Distinguish between various sampling methods
- Distinguish between types of variables
- Distinguish between null and alternative hypotheses
- Distinguish between directional and nondirectional hypotheses
- Write various types of hypotheses
- Understand the relationship between a researcher question and the directionality of the hypothesis
- Understand the impact of extraneous variables on interpretation of a study's result

Module 13

- Understand that probability, being uncertain, always includes error
- Distinguish between two types of error—Type 1 and Type 2

Module 14

- Understand the logic underlying the numerator and denominator in most inferential test statistics

Computational Exercises

1. A teacher is interested in obtaining feedback from his students regarding their opinion of the class. However, he has over 500 students in the class and so he decides to take a sample. To avoid bias, he decides to divide his class into sections based on their grades and sample 10 students from each section. What type of sampling strategy is the teacher using?

2. A researcher is interested in seeing how a new type of drug affects sleep in rats. The researcher gives each rat the drug and then monitors their sleep. What is the IV and what is the DV in this study?

3. The researcher from the previous question finds that the rats with the drug tend to sleep more than those without the drug. She decides to see if this sleep effect can be offset by the presentation of light. Every rat is given the drug and half are placed in a well-lit chamber, while the other half are placed in a dark chamber. What is the IV and what is the DV in this study?

4. A developmental psychologist is interested in how children's math skills develop as they age. The psychologist monitors the development of math skills in a random sample of children as they progress from first to sixth grade. What is the IV and what is the DV in this study?

5. For the study proposed in Question 3, what is the null and alternative (research) hypotheses?

6. A photographer believes that black-and-white pictures are preferred over color pictures in a popular magazine. The magazine creates an opinion poll for its readers to provide the photographer with an answer. What are the IV and DV for this study? What are the null and alternative hypotheses?

7. The results of a study by a magazine publisher indicate that the readers have a strong preference for black-and-white pictures over color pictures. However, after publishing an issue with almost all black-and-white photographs, the magazine receives a number of angry letters from subscribers that state they love color photographs. These letters far outnumber the number of responses that they received in the poll. What type of error did the magazine publishers make in their conclusion from the original poll?

8. A research team is investigating the effects of a neuropeptide on alcohol addiction. They develop a study in which they are able to find a local group of individuals with this peptide and a local group without the peptide. The researchers then determine the extent to which these individuals are addicted to alcohol. What are the IV and DV for this study? What are the null and alternative hypotheses? Is this a post hoc test? Why or why not?

9. What type of sampling strategy was used in the study discussed in Question 8?

10. It has been consistently shown that exposure therapy is an effective treatment for specific phobias. You conduct a research study examining the effect of exposure treatment on participants with acrophobia (fear of heights) and find that the participants do not improve. You run to your colleagues to tell them about your groundbreaking finding, but they disapprove. What is likely to be the argument against your finding?

11. A z score of 2.4 that is drawn from a population with a standard deviation of 14 corresponds to a raw score of 56. What is the expected value when drawing a raw score from this population?

12. Sarah received a grade of 1,540 on the SAT. If the mean for the SAT that particular year was 1,000 with a standard deviation of 200, what percentage of the SAT population scored below her?

13. Murray's parents are concerned that he is sleeping too much. They recently read an article stating that the average person should sleep 7.5 hr a night with a standard deviation of 1 hr. Murray sleeps about 9.5 hr a night. What percentage of the population sleeps more than Murray? Less than Murray? What would you tell his parents regarding his sleeping?

14. A team of scientists are investigating the side effects of a new muscle enhancer on rats. They anecdotally discover that one of the rats with the muscle enhancer has become increasingly aggressive. This rat has bitten other rats 10 times in an hour whereas the average rat bites only 3 times an hour with a standard deviation of 1.5 bites. What are the null and alternative hypotheses? What is the probability that another rat would bite more than this rat? Would you reject or retain the null hypothesis?

15. You are the teacher of a new women's study course. At the end of the term, one of your students states that he or she has missed 5 days of class and is upset that he or she has had his or her final grade lowered as a result of his or her absences. He or she argues that he or she shouldn't be penalized for his or her absences because the other students have many more absences. In going over your class roll, you notice that the average student missed 2 days of class with a standard deviation of 1.7. What are the null and alternative hypotheses of this study? Is this study directional or nondirectional? How would you be inclined to respond to the student based on this information?

Answers to Odd-Numbered Computational Exercises

1. This is a stratified sampling technique.

3. The IV is now the presence of light, and the DV is the amount of sleep.

5. Null hypothesis: the presence of light will not affect the amount of time spent sleeping by rats taking the drug. Alternative hypothesis: the presence of light will affect the amount of time spent sleeping by rats taking the drug.

7. The publishers committed a Type 1 error.

9. This was a convenience sample because they were able to select local people that had or did not have the neuropeptide.

11. The expected value would be the mean, which is 22.4.

13. Approximately 2% of the population sleeps more than Murray. Approximately 98% of the population sleeps less than Murray. I would tell them that he does appear to be sleeping a lot more than would be expected.

15. Null hypothesis: The student was not absent more than other students. Alternative hypothesis: The student was absent more than other students. This would be a directional study. Since the percentage of students missing more classes would be approximately 4%, I would state that this student's argument is wrong and that he or she should be penalized for missing class.

True/False Questions

1. You can expect the means of a sample and population to slightly differ, but not their standard deviations.
2. Joe decides to pick his softball team by selecting every other person available to play. This is a random sampling technique.
3. The goal of stratified sampling is to have the characteristics of your sample be similar to those of the population.
4. Convenience sampling is among the optimal methods of sampling.
5. Confounding variables are desirable in research studies.
6. You are interested in determining if a sports drink improves athletic performance. The hypothesis in this study would be directional.
7. A significant difference refers to any difference between two samples.
8. Type 1 errors refer to instances when the null hypothesis is incorrectly rejected.
9. It is possible to make both a Type 1 error and a Type 2 error with the same hypothesis test.
10. One possible cause of a Type 1 or 2 error is obtaining a sample that is unrepresentative of the population.
11. The numerator of the z score formula is making a comparison between an obtained value and an expected value.
12. If there is a difference between the obtained and expected value, then you always reject the null hypothesis.
13. If you obtain a very high z score (>4), then there is no chance that you have committed a Type 1 error.
14. The standardized random error provides a measure of comparison for the difference between the observed value and the expected value.
15. In retaining the null hypothesis, you may have committed a Type 2 error.

Answers to Odd-Numbered True/False Questions

1. False
3. True
5. False
7. False
9. False
11. True
13. False
15. True

Short-Answer Questions

1. Why are representative samples important when doing research? In general, what is the best method to use when trying to obtain a representative sample?

2. Why is convenience sampling not the optimal choice when choosing a sample strategy?

3. What are the potential causes of sampling error? How does this differ from the causes of a "real" (significant) difference between two groups?

4. Does the researcher always have control over the independent variable? What (if any) are the potential circumstances when the researcher may not have control over the independent variable?

5. What is the purpose of the null hypothesis?

6. A scientist is interested in determining how a new type of chemical affects engine parts. They place the chemical on 20 engines and compare how well these engines run in comparison with the 20 engines without the chemical. Would the hypotheses for this study be directional or nondirectional? Why or why not?

7. What is a Type 1 error? What are the circumstances that would lead one to make this type of error?

8. What is a Type 2 error? What are the circumstances that would lead one to make this type of error?

9. Why would it be impossible to make both a Type 1 error and a Type 2 error simultaneously?

10. Are all differences between two groups meaningful differences? How could you go about determining whether or not this difference is in fact meaningful?

11. How does the formula for a z score provide a template or prototype for all inferential statistics?

12. How does probability aid our decision of whether or not we should retain or reject the null hypothesis?

13. When can you be 100% certain when you reject or retain the null hypothesis?

14. Mark is trying out a new diet in hopes of losing weight. He decides to measure himself prior to the diet and then monitor his weight change while dieting. What are the IV and DV? What are the null and alternative hypotheses? Is this a nondirectional or a directional test?

15. Jessie is studying the effect of a new herbal supplement on memory. She gives some of her friends in her biology class the supplement. She then determines how well these friends remember information as compared with some of her other friends in a chemistry class. What is the primary IV and DV of this study? What are potential confounding/extraneous variables in this study?

Answers to Odd-Numbered Short-Answer Questions

1. Representative samples are necessary to draw valid conclusions about the population in which the samples originated. If the sample is not representative, the inferential

statistical techniques that are used may be incorrect. The optimal method to obtain a representative sample is random sampling.

3. Sampling error is potentially caused by random chance error that causes the sample statistics and population parameters to differ. This differs from "real" differences in that whereas sampling error is caused by chance, a real difference is caused by another variable (hopefully the IV).

5. The null hypothesis serves a statistical/probability purpose. It represents what should be found if there is no effect of the IV. The purpose of a research study is to find support for the alternative hypothesis.

7. A Type 1 error is incorrectly rejecting the null hypothesis. This occurs when you obtain samples that are very different but their differences occurred by chance, not because of the effect of another variable.

9. Type 1 errors can only be made when the null hypothesis is rejected, whereas Type 2 errors can only be made when the null hypothesis is retained. After an inferential test, you cannot both retain and reject the null hypothesis.

11. The z-score formula provides a template in that you are determining the difference between an observed value and an expected value $(X - M)$. Then you are comparing the difference between the observed and expected values to a standardized error, or the amount of difference you expected (s). If the difference between the values is similar to what was expected, then you can retain the null hypothesis. Otherwise, the null hypothesis is rejected.

13. You can never be 100% certain that you can reject or retain the null hypothesis. There is always the chance for an error.

15. The primary IV is the use of the supplement and the DV is the amount of information memorized. The potential confounding variable is class. The information in the biology class may be easier to memorize than that of the chemistry class (or vice versa). This could influence the results of the test.

Multiple-Choice Questions

1. Which type of sampling is most prone to give you the least representative sample?
    a. Random sampling
    b. Stratified random sampling
    c. Convenience sampling
    d. Cluster sampling

2. Representative samples should
    a. Include participants who have similar characteristics to those in the population
    b. Have sample statistics that are identical to population parameters
    c. Be carefully handpicked from the population
    d. Try to capture all the variability in the population

3. In a study of language acquisition in chimpanzees, chimpanzees are randomly assigned to one of two groups. The first group receives language training with symbols, while the other group just interacts with humans. At the end of 1 month, the language acquisition of the chimpanzees is reviewed. However, the researchers notice that the chimps in the language group were slightly younger than those in the other group. What is the independent variable in this study?

a. Language acquisition
b. Age
c. Time
d. Language training versus human interaction

4. Using the study from Question 3, what is a potential confounding variable?
   a. Language acquisition
   b. Age
   c. Time
   d. Group

5. A team of scientists are interested in decreasing the time it takes for cookies to bake. They use a new food supplement in the cookie recipe and watch the baking time for the cookies. What is the null hypothesis in this study?
   a. The food supplement will change the time it takes for cookies to bake
   b. The food supplement will not change the time it takes for cookies to bake
   c. The food supplement will decrease the time it takes for cookies to bake
   d. The food supplement will increase the time it takes for cookies to bake

6. Any alternative hypothesis will always state
   a. That the IV affected the DV in a specific direction
   b. That the IV did not affect the DV in a specific direction
   c. That the IV affected the DV
   d. That the DV affected the IV

7. Sampling error is not considered to be related to
   a. The effect of the IV on the DV
   b. Random chance error
   c. Slight deviations between the sample statistics and population parameters
   d. The size of the sample

8. A significant difference refers to
   a. A very large difference between two groups
   b. Any difference between two groups
   c. A difference between two groups that is greater than would be expected by chance
   d. A difference between two group means that is less than would be expected by chance

9. If you were to retain a true null hypothesis then you would
   a. Have committed a Type 1 error
   b. Have committed a Type 2 error
   c. Have made a correct decision
   d. The answer depends on the null and alternative hypotheses

10. If you were to reject a true null hypothesis then you would
    a. Have committed a Type 1 error
    b. Have committed a Type 2 error
    c. Have made a correct decision
    d. The answer depends on the null and alternative hypotheses

11. A Type 1 error can occur when
    a. Comparing two misrepresentative samples that were drawn from the same population and thus appear to be significantly different
    b. Comparing two representative samples that were drawn from the same population and thus appear to be similar

c. Comparing two representative samples that were drawn from different populations and thus appear to be significantly different
d. Comparing two misrepresentative samples that were drawn from different populations and thus appear to be similar

12. You can be certain that you have committed a Type 2 error when
    a. You have a very large difference between your groups and you expected a very small difference
    b. You have a very small difference between your groups and you expected a very large difference
    c. Other studies have clearly indicated that the IV does affect the DV
    d. You can never be certain that you have committed a Type 2 error

13. What does the denominator in the z-score formula represent?
    a. A comparison between an observed value and an expected value
    b. An expected amount of random error expected from observed values
    c. The probability of obtaining this score by chance
    d. The standardized score of your obtained score

14. The average test grade on a science test was 75 with a standard deviation of 6. If April received a grade of 82, what percentage of the class did she do better than?
    a. .80
    b. .82
    c. .88
    d. .95

15. Using the above example, what would be the alternative hypothesis if April wanted to determine if she scored significantly higher than the other students in her class?
    a. April did significantly better than other students
    b. April did significantly different from other students
    c. April did not do significantly better than other students
    d. April did not do significantly different from other students

16. The national average for an anxiety inventory is 30 with a standard deviation of 7. If you are a therapist and have been seeing a client who now reports a score of 24, what percentage of the population would score as low or lower than your client?
    a. 41%
    b. 32%
    c. 20%
    d. 15%

17. What is the best way to obtain a random sample?
    a. Obtain an alphabetical list of the population of interest and select every third name
    b. Assign everyone in the population of interest a random number and then randomly pick numbers yourself
    c. Assign everyone in the population of interest a random number and then have a computer program randomly pick numbers
    d. Assign everyone in your immediate area who is part of the population of interest a random number and then have a computer program randomly pick numbers

18. Sandra is completing a study on postpartum depression. She was interested in determining if the age at which mothers give birth has an effect on the length of postpartum depression. Which of the following best describes this study?
    a. Post hoc, nondirectional, IV: mother's age
    b. Directional, IV: mother's age

c. Nondirectional, DV: length of depression
d. Post hoc, directional, DV: length of depression

19. Bert is interested in determining if solar power will reduce his electric costs. He puts solar panels on his roof and then compares his power bill prior to the panels with his bill after the panels. Which of the following best describes this study?
    a. Post hoc, nondirectional, IV: solar panels
    b. Directional, IV: solar panels
    c. Nondirectional, DV: cost of power bill
    d. Post hoc, directional, DV: cost of power bill

20. Thomas is studying aggression. He believes that excessive stimulation will lead to an aggressive reaction in cats. To assess this, he divides a sample of cats into 2 groups: the first will receive continuous petting for 3 hr, and the second will receive a standard amount of petting for 3 hr. Which of the following best describes this study?
    a. Directional, IV: amount of petting, DV: aggressive reaction
    b. Directional, IV: aggressive reaction, DV: amount of petting
    c. Post hoc, nondirectional, DV: amount of petting
    d. Post hoc, directional, IV: amount of petting

Answers to Odd-Numbered Multiple-Choice Questions

1. c
3. d
5. b
7. a
9. c
11. a
13. b
15. a
17. c
19. b

Part VIII Study Guide

The One-Sample Test

Part VIII Summary

Module 15

- If you graph the means of an infinite number of samples of any size from any distribution, the graph would appear to be normally distributed. This is the case regardless of the shape of the original distribution. This principle is referred to as the *central limit theorem*. Also, as the size of each sample increases, the effect of the central limit theorem becomes more pronounced. In fact, samples with an $n > 30$ rarely deviate from the center of the distribution of sample means.
- The distribution that is formed by the sample means is referred to as the *sampling distribution of the mean*. The mean of the sampling distribution of the mean is always going to be equal to the population mean.
- The sampling distribution of the mean also contains a measure of the average deviation of a sample mean from the population mean. Since we are now dealing with sample statistics instead of raw scores, the term that refers to the average deviation of a sample mean from the population mean is referred to as a *standard error*. The standard error for a sampling distribution of the mean is called the *standard error of the mean*. The standard error of the mean is calculated by using the following formula:

$$\frac{\sigma}{\sqrt{n}} = \sigma_M$$

- The standard error of the mean will always be smaller than the standard deviation of the population. Furthermore, σ_M will become smaller as sample size increases. This is because as you increase the size of your sample, you are better able to approximate the population. In better approximating the population, you reduce the amount of sampling error you expect between your sample and the population. Conversely, σ_M will become larger as the sample size decreases until you have a sample of 1, in which case σ_M will be equal to σ.

Module 16

- The primary purpose of a hypothesis test is to determine if the difference between the sample mean and the population mean is substantial enough to believe that the sample comes from a different population (indicating that the IV had an effect). This logic is embodied in the formula for a Z test, or a *normal deviate Z test*. This formula determines the difference between the sample mean and the population mean in the numerator ($M - \mu$) and then compares this with the average amount a sample mean should deviate from the population mean, or the standard error of the mean. Thus, the formula appears as follows:

$$\frac{M - \mu}{\sigma_M} = Z_{\text{norm dev}}$$

- The result from this test will tell you how far the sample mean falls from the population mean in standard error units. You can then refer to the table in Appendix A to determine the percentage chance of obtaining a sample mean above or below the mean of your current sample. Based on this probability, you can determine the chances that you would have obtained the sample from the population in the comparison. In research, the population from which a sample is drawn is rarely known and this test can help infer about the population from which the sample was drawn. For example, you obtain a sample mean with a z score of two. This means that you would expect to draw a sample with this mean or higher 2.28% of the time from this population. In other words, the chances of drawing a sample mean from this population are so slim that this sample may be more representative of a different population.

Module 17

- A one-sample t test is very similar to a Z test in that it follows similar logic and uses a similar process in determining whether or not a sample is significantly different from a population. However, a t test is used in lieu of a Z test when either (1) the size of your sample is less than 25 or (2) you do not know the standard deviation for the population.
- The main distinction between the formula for a one-sample t test and a Z test is the denominator, or the standard error. In a Z test, the standard error was calculated by dividing the population standard deviation by the square of the sample size (n). In a one-sample t test, the population standard deviation is unknown. Therefore, you must use the *estimated population standard deviation,* which places $n - 1$ in the denominator. The formula for the estimated population standard deviation is as follows:

$$\sigma_{\text{est}} = \sqrt{\frac{\sum (X - M)^2}{n - 1}}$$

- The denominator of the estimated population standard deviation is also the degrees of freedom. *Degrees of freedom* refers to the number of values that are allowed to vary when calculating a statistic. For example, imagine that you have a sample with an $n = 3$ and an $M = 10$. Your first two scores could be any values that you choose, but the third value must make the overall mean of the three scores 10. Let's say that the first two scores were a 7 and a 4. This indicates that for the mean of this sample to be 10, the third value must be 19. For this example, there were 2 *df*, two scores that could have been any value. Thus, the formula for degrees of freedom for a one-sample t test is $n - 1$. This formula will change depending on the type of test you are conducting.
- The reason that $n - 1$ is used in the denominator when estimating the standard deviation of a population is to correct for bias. The variability of a sample will always be smaller than the variability of a population. This is because a sample will tend to miss many of the extreme scores, or outliers, in a population that increase the population

variability. To correct for this bias in the sample, we subtract 1 from the denominator to increase the value of the standard deviation, which will better approximate the population standard deviation. This $n - 1$ is called a *correction factor*.
- The null hypothesis is retained when the difference between the population mean and the sample mean is *close to zero*. In contrast, the null hypothesis is rejected when the difference between the population mean and sample is *far from zero*. This is determined by dividing the sampling distribution into two pieces. The piece that represents the area close to zero (or close to the mean) is the *region of retention*. The area that falls far from the mean, or in the tails, is referred to as the *region of rejection*.
- In a *nondirectional* test, the alternative hypothesis states that there is a difference between the two samples, but it does not specify in which direction the difference is expected. In other words, there could be a significant increase or a decrease. In contrast, a *directional* test does indicate a specific direction; the sample mean is expected to be higher or lower than the population mean. This is important for determining where the regions of retention and rejection are located. For a directional test, the region of rejection is placed entirely in one tail. This type of test is referred to as a *one-tailed* test. For a nondirectional test, the region of rejection is placed in both tails. This type of test is referred to as a *two-tailed* test.
- After determining where the region of rejection will fall (one or two tails), you should decide the level of Type 1 error that you are willing to accept. This refers to the chance that the null hypothesis may be incorrectly rejected.
- Another aspect of note for the one-sample t test is the shape of the distribution that is used to determine the probability of obtaining this particular sample mean. In a Z test, the shape of the distribution will always be the same so the probabilities of that curve area will always be the same. However, because we are using estimates (sample statistics) in our inferential statistics, the shape of the distribution must be amended. We can expect that at smaller sample sizes, the shape of the distribution will deviate further from normality. Thus, the t distributions, because they are actually a family of distributions, have slightly raised tails and appear somewhat leptokurtic. However, as the sample size increases past 30, the t distributions that are used are practically identical to the normal curve.
- The scores that mark the region of rejection, or the *critical values*, for the t distributions can be found in Appendix C. The table is used as follows. First, find the degree of freedom that corresponds to the sample size. If the sample size is not listed, then use the next lowest sample size. Then, find the column that corresponds to the error rate. Be mindful that the table provides error rates for one-tailed and two-tailed tests. Then, find the t value that corresponds to your *df* and predetermined Type 1 error rate. If the t value you have obtained from your t test exceeds the value in the table, then the null hypothesis is rejected. If the t value you have obtained does not exceed the value on the table, the null hypothesis is retained.

Module 18

- In statistics, *confidence* refers to the probability that a Type 1 error was *not* made. In other words, it's the chance that we have found a "real" effect (the changes in our DV are really caused by our IV). Using the tables, it is impossible to determine the actual confidence for our hypothesis test. This is because the table does not provide the specific probability, or p, of actual incurred error. Using p, confidence can be expressed as $1 - p$.
- It is common to express your results in APA format, which appears as follows for a one-sample t test: $t(\text{df}) = t$ test value, $p = p$ value.

- The decision to reject or retain the null hypothesis is based on where the test statistic (value obtained from your *t* test) falls on the curve, in the region of rejection or retention. Once again, these areas are set by the predetermined α, which is conventionally set at .05. The use of α makes rejecting or retaining the null hypothesis a dichotomous decision, which can be problematic. The α level that is used can determine the results of a study, which can have larger implications. It may appear unfair that there is no "middle ground" decision in this matter. As such, you should not consider the rejection or retention of the null hypothesis as a divine message. Rather, you should view it as support for one hypothesis but be open to the possibility that these results may differ with another sample. However, you can feel more confident in your decision as the *p* value decreases (as this means your chances of making a Type 1 error is very small).
- *Parameter estimations* are estimations of population parameters that are based on sample statistics. There are two types of parameter estimates. The first is referred to as a point estimate. *Point estimates* are made when using a sample statistic to make a single guess about a population parameter. Stating that a population mean is 7 is a point estimate. The second parameter estimation is an interval estimate. *Interval estimates* suggest that the population parameter falls within a range of values.
- *Confidence intervals* are interval estimates that provide a range of values in which the population parameter is expected to fall. The sizes of these intervals are determined by a set percentage. This percentage refers to the number of times you would expect the population parameter to fall within the interval if the study were done repeatedly. For example, a 75% confidence interval means that 75% of the time, you would expect the population parameter to fall within the interval and 25% of the time, the parameter would not fall within the interval. Confidence intervals are usually estimated at 95% and 99%. The formula for a confidence interval is as follows:

$$CI = M \pm (t_{crit\ at\ .5\ \alpha})(\sigma_M)$$

Learning Objectives

Module 15

- Understand the difference between a raw score distribution and a sampling distribution of the mean
- Understand why any sampling distribution of the mean is normally distributed
- Understand the impact of sample size on the shape of a sampling distribution
- Understand the impact of sample size on the size of the standard error of the mean

Module 16

- Distinguish between a *z* score and a *Z* test
- Know the conditions under which it is appropriate to use a normal deviate *Z* test
- Calculate a normal deviate *Z* test
- Use a normal curve table to interpret a normal deviate *Z*

Module 17

- Distinguish between a normal deviate *Z* test and a one-sample *t* test
- Know the conditions under which it is appropriate to use a one-sample *t* test

- Understand the similar logic underlying various test statistics
- Understand the concept of degrees of freedom
- Distinguish between biased and unbiased estimators of a population parameter
- Find the regions of retention and rejection
- Understand the relationship between directionality of the hypothesis and the tail in the sampling distribution
- Calculate a one-sample t test
- Use a table to interpret calculated t
- Report results in APA format

Module 18

- Distinguish between tabled and incurred alpha
- Understand the relationship between error and confidence
- Estimate parameters—point and interval

Computational Exercises

1. Samples of $n = 16$ are drawn from a population with $\mu = 50$ and $\sigma = 10$. What are the μ and the standard error of the sampling distribution of the mean?

2. Using the information from Question 1, change the sample size to $n = 24$. Now what are the μ and the standard error of the sampling distribution of the mean? What has changed between the two samples and why?

3. The mean of a population is 25 and the standard deviation is 4. If you were to draw an infinite number of samples with $n = 12$, what sample mean would you expect to find two standard errors above the mean?

4. Using the normal distribution and a nondirectional test, what would your critical value be for a Type 1 error rate (α) of .05? .01? .1? .3?

5. The principal of a high school is seeking to improve his school's performance on a foreign language test by recruiting a new teacher for one of the classes. Last year, the school had a $\mu = 83$ and $\sigma = 5$. The new class has $n = 32$ students and obtains a mean grade of 87 on this year's test. Determine if this teacher has been able to improve his students' performance on the test.
   a. What are your hypotheses?
   b. Is this a directional or nondirectional test?
   c. Determine whether you should reject or retain the null hypothesis at the $\alpha = .05$ level.
   d. What error would you have potentially made in your decision?

6. The owner of a retirement community wants to improve the satisfaction of the residents by implementing a new exercise activity. The prior satisfaction, as rated by a 0 to 10 scale, was $\mu = 6$ and $\sigma = 1.2$. Only $n = 12$ members participate in the new activity and then rate their satisfaction as an $M = 8$. Did the new activity improve satisfaction ratings?
   a. What are your hypotheses?
   b. Is this a directional or nondirectional test?
   c. Determine whether you should reject or retain the null hypotheses at $\alpha = .01$.
   d. What error would you have potentially made in your decision?

7. An appliance company is trying to improve the strength of their blow-dryer. Their old model would dry hair in μ = 130 s with a σ = 22. The new blow-dryer was able to dry a sample of $n = 12$ people's hair in $M = 120$ s. What is the probability that one of the old blow-dryers would be able to dry hair as fast or faster than this newer model?

8. A fashion designer wants to determine the effect that changing the color of her headbands would have on their sales. The headbands currently sell μ = 8 and σ = 1 per week in stores across the nation. In a conservative move, the designer releases the new color headbands to only $n = 4$ stores and monitors their sales for the week. The new headbands sell with an $M = 10$ during the week.
   a. What are your hypotheses?
   b. Is this a directional or nondirectional test?
   c. Determine whether you should reject or retain the null hypothesis at α = .05.
   d. What error would you have potentially made in your decision?

9. You notice an advertisement for a speed-reading course that claims that it vastly improves reading rate. It stated that the average reading rate of μ = 50 words per minute in a class of $n = 30$ improved to 65 words per minute. You are skeptical of this advertisement and are able to track down all 30 participants and retest them. Sure enough, you obtain a sample mean of 65 but a standard deviation of 41. Is this course really able to improve reading rate?
   a. What are your hypotheses?
   b. Is this a directional or nondirectional test?
   c. Determine whether you should reject or retain the null hypothesis at α = .05.
   d. What error would you have potentially made in your decision?

10. A farmer is curious about the effect that new plant food will have on her tomato crop. Previously, the tomatoes she grew were μ = 0.8 lbs. She uses the new plant food and grows $n = 13$ tomatoes that weigh an $M = 1.1$ lbs with an $s = 0.7$. What is the probability that tomatoes of this weight or lower would grow under the null hypothesis? Is it likely that she would obtain this crop?

11. A national anxiety survey finds that individuals with generalized anxiety disorder (GAD) score at approximately μ = 13 on an anxiety measure. You are developing a new treatment to address GAD and implement it in a sample of $n = 6$ participants. You obtain an $M = 11$ and an $s = 1.3$ from this sample. Is your treatment effective at reducing anxiety?
    a. What are your hypotheses?
    b. Is this a directional or nondirectional test?
    c. Determine whether you should reject or retain the null hypothesis at α = .05.
    d. What error would you have potentially made in your decision?

12. A researcher has developed a treatment for post-traumatic stress disorder (PTSD) that is designed to be much shorter than prior interventions. Currently, PTSD treatment usually lasts for μ = 8 sessions. The new treatment has been used with $n = 4$ participants and has shown results in $M = 6$, $s = 2$ sessions. Is this new treatment significantly briefer than regular treatments?
    a. What are your hypotheses?
    b. Is this a directional or nondirectional test?
    c. Determine whether you should reject or retain the null hypothesis at α = .01.
    d. What error would you have potentially made in your decision?

13. Using the information in Question 9, find a 95% confidence interval.

14. Using the information in Question 10, find a 99% confidence interval.

15. Using the information in Question 12, find a 95% confidence interval.

Part VIII Study Guide

Answers to Odd-Numbered Computational Exercises

1. $\mu = 50$; $\sigma_M = \dfrac{10}{\sqrt{16}} = 2.5$

3. $\mu = 25$; $\sigma_M = \dfrac{4}{\sqrt{12}} = 1.15$. Two standard errors above the mean would be a mean of 27.3.

5.
   a. Null hypothesis: The new teacher will not improve the performance of his students. Alternative hypothesis: The new teacher will improve the performance of his students.
   b. This is a directional test (seeking to improve performance).
   c. $= \dfrac{5}{\sqrt{32}} = 0.88$; $Z = \dfrac{87 - 83}{0.88} = 4.55$; reject the null hypothesis
   d. Type 1 error

7. $\sigma_M = \dfrac{22}{\sqrt{12}} = 6.36$; $Z = \dfrac{120 - 130}{6.36} = 1.57$; there is approximately a 6% chance that another blow-dryer would dry hair this quickly.

9.
   a. Null hypothesis: The course will not improve reading rate. Alternative hypothesis: The course will improve reading rate.
   b. This is a directional test.
   c. $\sigma_M = \dfrac{41}{\sqrt{30}} = 7.49$; $Z = \dfrac{65 - 50}{7.49} = 2$; reject the null hypothesis
   d. Type 1 error

11.
   a. Null hypothesis: The treatment will not reduce the scores of individuals with GAD. Alternative hypothesis: The treatment will reduce the scores in individuals with GAD.
   b. This is a directional test.
   c. $= \dfrac{1.3}{\sqrt{6}} = 0.53$; $Z = \dfrac{11 - 13}{0.53} = -3.77$; reject the null hypothesis
   d. Type 1 error

13. CI = 1.1 ± (2.179)(0.19), 0.69 to 1.51

15. CI = 6 ± (3.182)(1), 2.82 to 9.18

True/False Questions

1. The central limit theorem states that the distribution of sample means will be normal, but only if the underlying distribution is skewed.

2. The sampling distribution of the mean is made up of the means of all the possible samples of a particular sample size.

3. With a sample size of $n = 1$, the standard error will be equal to the standard deviation.

4. Increasing sample size will decrease the standard error.

5. A score that falls 2 standard deviations above the mean falls in a different location on the normal curve than a mean that falls two standard errors above the mean.

6. In the normal deviate (Z) test, a smaller standard error means that the distribution appears more leptokurtic.

7. The normal deviate Z test will tell you if your sample came from a specific population.

8. Determining whether you are correct in rejecting or retaining the null hypothesis is never perfectly apparent.

9. A one-sample t test differs from a Z test in that it uses the estimated population standard deviation.

10. The estimated population standard deviation is a biased estimator in that it will always overestimate the population standard deviation.

11. The null hypothesis refers to the question, "Is the difference between the sample mean and the population mean close to zero?"

12. The region of retention is the portion of the curve that is in the tails.

13. *Directional hypothesis* specifies that the sample mean is expected to be either above or below the population mean.

14. The t distribution is actually a family of distributions as opposed to a single distribution.

15. The size of the confidence interval increases as the percentage of confidence increases. In other words, a 95% confidence interval is smaller than a 99% confidence interval.

Answers to Odd-Numbered True/False Questions

1. False
3. True
5. True
7. False
9. True
11. True
13. True
15. True

Short-Answer Questions

1. What is the central limit theorem? Why is it important to inferential statistics?
2. Why is the mean of the sampling distribution of the mean always equal to the population mean?

3. Why does the standard error of the mean shrink as the size of your sample increases?
4. What pieces of information does a z score from a normal deviate test provide?
5. How are statistical decisions made using a normal deviate Z test?
6. When would you use a one-sample t test instead of a normal deviate test?
7. How does the estimated population standard deviation differ from the formula for the population deviation? Why is this change made to the formula?
8. Why is the estimated population standard deviation a biased indicator?
9. Compare and contrast the region of rejection for a nondirectional and directional test.
10. How do the t distributions differ from the standard normal distribution? How are the t distributions affected by sample size?
11. Why would a researcher move his or her α level from .05 to .01?
12. If a result is written as $p < .05$, can you determine the exact level of confidence? Why or why not?
13. How are confidence intervals affected by the size of the sample?
14. How does the size of the confidence interval change as you decrease the level of confidence (i.e., moving from a 99% to a 95% CI)?
15. What is a confidence interval?

Answers to Odd-Numbered Short-Answer Questions

1. The central limit theorem states that the distribution that results from drawing an infinite number of samples of the same size from any population will be normal. This is important to inferential statistics because it allows the proportions of the normal curve to be used for hypothesis testing.

3. As the sample size increases, it is better able to approximate the variability of the population. As the sample becomes a more accurate representation of the population, there is less expected deviation between the samples and the populations.

5. First, the distance of the sample mean from population mean is compared with the standard error, the expected deviation. Then, the probability of obtaining such a sample mean is determined by using the z table. If the probability of obtaining this mean is less than the critical value (or value that denotes the allowed Type 1 error level), then the null hypothesis is retained. However, if the probability of obtaining the sample mean is greater than the critical value, the null hypothesis is rejected.

7. The difference is in the denominator, which for the standard deviation contains only n, whereas the estimated population standard deviation contains $n - 1$. This change is used to correct the bias that is present in the estimated population standard deviation.

9. In a directional test, the region of rejection is located entirely in one tail. In a nondirectional test, the region of rejection is divided into two equal parts that are located in each of the tails.

11. This would be done to reduce the chances of making a Type 1 error. This sets up more stringent criteria for rejecting the null hypothesis.

13. Confidence intervals are decreased as sample size increases. This is because larger samples will result in a smaller critical value and a smaller standard error. As a result, a larger sample will decrease the standard error, reducing the size of the interval.

15. A confidence interval provides a range of values within which the sample mean would probably fall if the study were replicated numerous times.

Multiple-Choice Questions

1. According to the central limit theorem, if a population is very positively skewed, the resulting distribution that would be created by samples of $n = 45$ will be
   a. Normally distributed
   b. Positively skewed
   c. Negatively skewed
   d. Bimodal

2. The standard error will almost always be _____ when compared with the standard deviation
   a. Larger
   b. Smaller
   c. Equal
   d. It depends on the sample size

3. The standard error for a distribution of samples of $n = 1$ would be
   a. 1
   b. 2
   c. 3
   d. σ

4. For a population with $\sigma = 6$, what would be the standard error for samples of 36?
   a. 6
   b. 36
   c. 1
   d. 54

5. For a population with $\sigma = 10$, what would be the standard error for samples of 45?
   a. 1.49
   b. 2
   c. 3.41
   d. 4.32

6. The grades on a Spanish test in a foreign language class for the past 5 years have been an average of $\mu = 75$ with a $\sigma = 4$. This year, Simon received a grade of 80 on the test, which was the average of his class of $n = 30$. What is the probability that someone did better than Simon on this test?
   a. .75
   b. .68
   c. .32
   d. .11

Part VIII Study Guide

7. Using the information from Question 6, what is the probability that the class from another year would do better than Simon's class?
   a. .10
   b. .00
   c. .11
   d. .31

8. A developmental psychologist is studying motor development by examining the ages at which children are able to take 3 steps. If the average child can take 3 steps at age $\mu = 1.2$ years with a $\sigma = 0.3$ years, what is the probability of finding a group of $n = 5$ children who could walk at age 0.9 years?
   a. .16
   b. .99
   c. .01
   d. .32

9. Using the information from Question 8, what is the probability of finding a group of $n = 12$ children that could walk at a mean age of 1.3 years or greater?
   a. .3
   b. .01
   c. .00
   d. .46

10. Using the responses to Questions 8 and 9, for which scenario would you feel more comfortable rejecting the null hypothesis?
    a. Question 8
    b. Question 9
    c. Questions 8 and 9
    d. Neither

11. In which way is a sample standard deviation a biased estimate of the population standard deviation?
    a. It underestimates the population variability
    b. It overestimates the population variability
    c. The direction of the bias depends on the sample size
    d. There is no bias if the sample size is less than 10

12. If you are conducting a nondirectional Z test with α of .01, what is the area of the region of rejection in each tail?
    a. .01
    b. .005
    c. .02
    d. .10

13. As sample size decreases, how does the t distribution change?
    a. It approaches the normal distribution
    b. It becomes more platokurtic
    c. The tails become more elevated from the number line
    d. b and c

14. Tessa is studying the behavior of beetles. She wants to determine the effect of a new neurochemical on the number of eggs a female beetle lays. Without the neurochemical, beetles lay $\mu = 14$ eggs. With the neurochemical, the $n = 25$ beetles lay

$M = 15.3$ eggs with an $s = 2.5$. What should Tessa conclude about the effect of the neurochemical if she is willing to accept a 5% Type 1 error rate?
a. Reject the null hypothesis of a nondirectional test
b. Retain the null hypothesis of a directional test
c. Reject the null hypothesis of a directional test
d. Retain the null hypothesis of a nondirectional test

15. Sari is a huge fan of her college's basketball team. However, her team has lost the past few games, and she is determined to figure out the cause of this streak. She hypothesizes that her team has shorter players than the rest of the league. Using the college basketball database, she determines that the μ height is 87 in. The mean height of her team with $n = 12$ players is 85.3 with an $s = 3.2$. If Sari is willing to accept an error rate of 1%, what should she conclude?
a. Reject the null hypothesis of a nondirectional test
b. Retain the null hypothesis of a directional test
c. Reject the null hypothesis of a directional test
d. Retain the null hypothesis of a nondirectional test

16. A guitar typically retails for $\mu = \$200$, $\sigma = \$25$. A popular online auction Web site has the guitar available from $n = 8$ sellers who are selling the guitar for $M = \$189$. If you want to purchase eight guitars for your music shop, what is the probability that you would find another sample of eight guitars for less?
a. 40%
b. 11%
c. 23%
d. 89%

17. A shoe company is trying to improve the walking comfort of its high heels. The results from previous tests have rated the comfort of the heels at $\mu = 45$ on a scale of 0 to 100. The company has created a new shoe and gives it to a sample of $n = 8$ to provide comfort ratings. If the new sample has a rating of $M = 50$ with an $s = 16$, what should the company conclude about the new shoes?
a. Reject the null hypothesis of a nondirectional test
b. Retain the null hypothesis of a directional test
c. Reject the null hypothesis of a directional test
d. Retain the null hypothesis of a nondirectional test

18. What is the 95% confidence interval for a nondirectional hypothesis test in which $M = 13$, $s = 4$, $n = 36$, and $\alpha = .05$?
a. 11 to 17
b. 25.57 to 28.25
c. 11.63 to 14.37
d. 10.71 to 17.59

19. What is the 99% confidence interval for a nondirectional hypothesis test in which $M = 24$, $s = 2.7$, $n = 10$, and $\alpha = .01$?
a. 21.23 to 26.77
b. 30.45 to 37.25
c. 21.3 to 26.7
d. 30 to 35

20. Rosa is very nervous about taking the GRE next week. She does a lot of preparation for the test and has taken $n = 7$ practice tests and obtained an $M = 1,270$ with $s = 45$. The population mean for the GRE is 1,120. Calm Rosa's nerves by providing her

with the 99% confidence interval (nondirectional) for her performance on the practice GREs that will help give her an idea of how well she will do on the actual test.
   a. 1,225 to 1,315
   b. 1,075 to 1,165
   c. 1,056.94 to 1,183.06
   d. 1,206.94 to 1,333.06

Answers to Odd-Numbered Multiple-Choice Questions

1. a
3. d
5. a
7. b
9. d
11. a
13. d
15. b
17. b
19. a

Part IX Study Guide

The Two-Sample Test

Part IX Summary

Module 19

- In two-sample research, we compare two samples that are thought to be representative of separate populations. This comparison enables us to answer the question, "Are these populations significantly different from one another?"
- The null hypothesis for two-sample research is that there *is no* difference between the populations (any difference between the samples is due to sampling error). In contrast, the alternative hypothesis for two-sample research indicates that there *is* a difference between the populations.
- The distribution for two-sample research is called the *sampling distribution of the difference between the means*. It is created by subtracting the means of all possible sample pairs from the populations that are being compared. Since the null hypothesis states that there is no difference between the populations, we can assume that the most common mean difference between these pairs of samples will be 0. Also, if the null hypothesis is true, we can assume that there is a low probability of obtaining a very large difference between sample means. This indicates that the sampling distribution of the difference between the means will be normally (symmetrically) distributed with a mean of 0.
- The standard deviation of the sampling distribution of the difference between the means is referred to as the *standard error of the difference between the means* ($\sigma_{M_1 - M_2}$). It is computed by *pooling* (combining) the variance of the two samples that are being compared. When the sample sizes are (1) equal and the samples are (2) independent, you can use the *special-case formula* for the standard error of the difference between the means. This formula is

$$\sigma_{M_1 - M_2} = \sqrt{\sigma_{M_1}^2 + \sigma_{M_2}^2}$$

- The standard error of the difference between the means is critical for determining if the difference between two samples is a significant or a nonsignificant difference. Larger sample sizes will reduce the size of $\sigma_{M_1 - M_2}$ whereas smaller sample sizes will increase the size of $\sigma_{M_1 - M_2}$. A smaller $\sigma_{M_1 - M_2}$ will increase the chances of finding a significant difference between two samples, whereas a larger $\sigma_{M_1 - M_2}$ will decrease the chances of finding a significant difference.

Module 20

- The format of the two-sample t test is similar to the other hypothesis tests that we have covered. It compares "what you got" (the difference between two sample means; $M_1 - M_2$) with "what you expected" (the expected difference between the two populations; $\mu_1 - \mu_2$), divided by the "standardized random error" (the standard error of the difference between the means; $\sigma_{M_1 - M_2}$).

$$t_{2\text{-samp}} = \frac{(M_1 - M_2) - (\mu_1 - \mu_2)}{\sigma_{M_1 - M_2}}$$

- The expected difference between the two populations ($\mu_1 - \mu_2$) under the null hypothesis is always expected to be zero. This reduces the formula to

$$t_{2\text{-samp}} = \frac{(M_1 - M_2)}{\sigma_{M_1 - M_2}}$$

- The table found in Appendix C contains the t table used for determining if the difference between the sample means is *only a little* (attributed to sampling error) or *a lot* (attributed to the independent variable). As in previous hypothesis tests, the means are considered significantly different if the value obtained from the t test is greater than the value obtained from the table (critical t). Critical t is obtained by using the degrees of freedom, for a two-sample test, which is found by using the following formula: $df = (n - 1) + (n - 1)$. This formula could also be rewritten as $df = N - 2$.
- After conducting any two-sample t test, the results are usually reported in APA format; t(degrees of freedom) = t_{obtained} value, $p <$ or $> \alpha$. For example, a $t_{\text{obtained}} = 3.5$ with a df of 4 that is significant at $\alpha = .05$ would be written as $t(4) = 3.5, p < .05$.

Module 21

- When comparing samples that have unequal sample sizes and are independent of each other you must use the *generalized formula* for $\sigma_{M_1 - M_2}$. This formula uses a process called *weighting* to account for difference in sample sizes when pooling the variances of samples with unequal ns. The generalized formula is

$$\sigma_{M_1 - M_2} = \sqrt{\left(\frac{SS_1 + SS_2}{n_1 + n_2 - 2}\right)\left(\frac{1}{n_1} + \frac{1}{n_2}\right)}$$

- The formula for a t test with unequal sample sizes then becomes

$$t_{2\text{-samp}} = \frac{M_1 - M_2}{\sqrt{\left(\frac{SS_1 + SS_2}{n_1 + n_2 - 2}\right)\left(\frac{1}{n_1} + \frac{1}{n_2}\right)}}$$

- The method we use to determine if the samples are significantly different is the same as that for a t test with equal sample sizes. The degrees of freedom are calculated with the following formula: $df = (n_1 - 1) + (n_2 - 1)$.

Module 22

- When the subjects in two samples are not independent, the study is called a *related-samples* study. There are two types of related-samples studies. The first is called *repeated measures*, which uses the same participants in both samples. The second is called *matched samples*, which uses different participants in each sample, but matches the participants in each sample on an extraneous variable. The benefit of a related-samples study is that it helps reduce the influence of the confounding variables that occur when you compare samples containing different people. Examples of these confounding variables could be level of education, socioeconomic status, and so on.

- Since the participants are related in some manner in a related-samples study, the responses in one sample are expected to be associated with the responses in the other sample. This relationship between the responses needs to be accounted for when finding t. This is done by removing the *covariance* between the groups in the calculation of $\sigma_{M_1 - M_2}$. Covariance is a measure of the extent to which two sets of scores vary together. The formula for $\sigma_{M_1 - M_2}$ in a related-samples t test is

$$\sigma_{M_1 - M_2} = \sqrt{\sigma_{M_1}^2 + \sigma_{M_2}^2 - 2r\sigma_1\sigma_2}$$

- The formula for a related-samples t test then becomes

$$t_{2\text{-samp}} = \frac{M_1 - M_2}{\sqrt{\sigma_{M_1}^2 + \sigma_{M_2}^2 - 2r\sigma_1\sigma_2}}$$

- Another method to calculate a related-samples t test uses the *direct-difference formula* (also called the *computational formula*). This formula allows you to do a related-samples t test without having to find r. In this formula, $\overline{D}$ represents the difference between all the paired scores:

$$t_{2\text{-samp}} = \frac{\overline{D} - \mu_D}{\sqrt{\dfrac{\sum D^2 - \dfrac{(\sum D)^2}{n}}{n(n-1)}}}$$

- The degrees of freedom for a related-samples t test is calculated by subtracting 1 from the number of *pairs* in the samples: $df = N - 1$.

Module 23

- The probability that we have not made a Type I error is called *confidence*. Confidence is calculated as $1 - \alpha$. Recall that α is the chance that we have made a Type I error. The precise level of α that is reported in a study is usually referred to as p. So actual confidence is calculated as $1 - p$.
- The confidence value is used in determining whether to retain or reject the null hypothesis. The probability that there is no significant difference between the samples (supporting the null hypothesis) is thought of as p. The probability that there is a

significant difference between the samples (rejecting the null hypothesis) is thought of as our confidence, $1 - p$.
- The social sciences will usually reject the null hypothesis when probability of making a Type I error (p) is less than .05. However, this rule is not set in stone. When reporting your results, you are encouraged to report the actual p value rather than stating that your p value was below or above .05. This allows the readers to make their own decision about the difference between the two means rather than making a dichotomous (significantly different/not significantly different) decision.
- After rejecting the null hypothesis, you can obtain an estimate of the range of the difference between the population means. This estimate is called the *parameter estimation*. There are two types of parameter estimates. The first is called a *point estimate*, in which the estimation of the difference between the two means is a single number. For two-sample research, the point estimate is usually the actual difference between the sample means. The second is called an *interval estimate*, which estimates a range in which the true population difference most likely falls.
- The range established by the interval estimate is also referred to as a *confidence interval*. Confidence intervals have an associated percentage that presents the confidence that the actual population mean difference falls within that range. For example, a 95% confidence of 5 to 10 means that you are 95% confident that the difference between two populations falls between the values of 5 and 10. The formula for calculating a confidence interval is

$$CI = (M_1 - M_2) \pm (t_{\text{critical at } 1/2\, \alpha})(\sigma_{M_1 - M_2})$$

Learning Objectives

Module 19

- Understand the difference between a sampling distribution of the mean and the sampling distribution of the difference between the means
- Understand why the mean of the sampling distribution of the difference between the means is zero
- Understand the impact of sample size on the size of the standard error of the difference between the means
- Understand the impact of the size of the standard error on the difference between the means of the test statistic t

Module 20

- Understand the similar logic underlying various test statistics
- Determine degrees of freedom
- Calculate a two-sample t test for independent samples and equal sample sizes
- Use a table to interpret calculated t
- Report results using APA format

Module 21

- Determine the degrees of freedom
- Calculate a two-sample t test for unequal sample sizes

- Use a table to interpret calculated t
- Report results in APA format

Module 22

- Distinguish between independent and related samples
- Determine degrees of freedom
- Calculate a two-sample t test for related samples
- Use a table to interpret calculated t
- Report results in APA format

Module 23

- Distinguish between tabled and incurred alpha
- Understand the relationship between error and confidence
- Estimate parameters—point and interval

Computational Exercises

1. What would be the standard error of the difference between the means for two samples in which $\sigma_1 = 7.5$ and $\sigma_2 = 4.27$?

2. What is the standard error of the mean for a study that is comparing the amount of sleep had by hibernating brown bears, $n = 10$, $\sigma_1 = 12.7$, with hibernating black bears, $n = 10$, $\sigma_2 = 5.23$?

3. The researchers in Question 2 were able to include an additional black bear, making $n = 11$ for this group. What is their new standard error for black bears?

4. You are interested in comparing the levels of stress between graduate students in a psychology program with graduate students in a business program. You give them a scale with range of 0 (no stress) to 7 (extremely stressed) and obtain the following data:

Psychology	Business
1	6
5	6
0	2
4	3
2	1
7	7

    a. What are the independent and dependent variables for this study?
    b. What type of t test should be used?
    c. State the null hypothesis and research hypothesis.
    d. Can you reject the null hypothesis at $\alpha = .05$? Report your findings in APA format.

5. A computer company is interested in seeing if monitor size *increases* the amount of time someone spends on the computer per day. They give a sample of individuals a 24-in. monitor and another sample of different people a 15-in. monitor and ask

them to record how many hours per day they use their computer. They obtain the following scores:

24 in.	15 in.
7	1
8	2
11	3
9	4
6	2
7	3
8	3
5	3
9	2

    a. What are the independent and dependent variables for this study?
    b. What type of t test should be used?
    c. State the null hypothesis and research hypothesis.
    d. Can you reject the null hypothesis at $\sigma = .05$? Report your findings in APA format.

6. Dr. Smith and Dr. Martinez are comparing the success rates of their different treatments for lung cancer. They are examining success as based on how long it takes for the patient to no longer have lung cancer. Here are their data:

Dr. Smith	Dr. Martinez
$n = 7$	$n = 9$
$M = 35.27$	$M = 32.47$
$SS = 167$	$SS = 178$

    a. What type of t test should be used?
    b. State the null hypothesis and research hypothesis.
    c. Can you reject the null hypothesis at $\alpha = .05$? Report your findings in APA format.

7. You notice that fruit bats and vampire bats are able to fly for different lengths of time. You hypothesize that this difference is attributed to a possible difference in their wing span. You conduct a study to test this hypothesis and obtain the following data.

Fruit Bats	Vampire Bats
$n = 4$	$n = 8$
$M = 8.27$	$M = 10.47$
$SS = 217$	$SS = 258$

    a. What type of t test should be used?
    b. State the null hypothesis and research hypothesis.
    c. Can you reject the null hypothesis at $\alpha = .05$? Report your findings in APA format.

8. College A recently beat College B in a weight lifting competition. The results of the match were very close with College A's $n = 6$ team members lifting an $M = 130$ pounds with an $s = 6$, and College B's $n = 6$ team members lifting an $M = 128$ pounds with an $s = 7$. The coach of College B's team decides to use some statistics to help improve his team's spirits. What could he tell them using $\alpha = .05$?

9. Dr. Johnson is interested in determining if his popularity with students is related to the time of his class. He wants to compare how much his students like him, as assessed by a 7-point rating scale (1 = *dislike* to 7 = *like*), in his 8:30 a.m. class and his 1:00 p.m. class. However, due to the very early time, there are fewer students in his 8:30 a.m. class. He obtains the following data:

8:30 a.m.	1:00 p.m.
$n = 16$	$n = 24$
$M = 3.4$	$M = 5.7$
$SS = 89$	$SS = 95$

   a. What are the independent and dependent variables for this study?
   b. What type of *t* test should be used?
   c. State the null hypothesis and research hypothesis.
   d. Can you reject the null hypothesis at $\alpha = .05$? Report your findings in APA format.

10. You are hired by a dental company to evaluate the impact of a new teeth-whitening process. The company informs you that they asked a sample of individuals to rate the whiteness of their teeth on a scale of 0 to 100. Then, the participants underwent the whitening process and were asked to rate the whiteness of their teeth again. They would like you to determine if the ratings significantly *improved*. Here are their data:

Before	After
45	57
82	94
67	76
21	64
51	50
32	48
37	47

   a. What are the independent and dependent variables for this study?
   b. What type of *t* test should be used?
   c. State the null hypothesis and research hypothesis.
   d. Can you reject the null hypothesis at $\alpha = .05$? Report your findings in APA format.

11. In a recent study of newlyweds, $n = 4$ married couples were asked to rate their marital satisfaction (on a −10 to 10 scale) after 2 years of marriage. The purpose of the study was to determine if couples differ on their perception of their marriage, so husbands and wives were asked independently, but the responses were matched across the couples. Here are the data:

Husband	Wife
10	9
8	7
2	4
4	6

   a. What are the independent and dependent variables for this study?
   b. What type of *t* test should be used?

c. State the null hypothesis and research hypothesis.
d. Can you reject the null hypothesis at $\alpha = .01$? Report your findings in APA format.

12. As the new head of a clothing company, you are asked to determine if women have a preference for skirts or pants. You collect a sample of $n = 7$ women and ask them to rate their preference for skirts and pants on a 1 to 5 scale. Here are the data:

Skirts	Pants
5	2
3	5
1	3
4	5
2	1
3	4

a. What type of t test should be used?
b. State the null hypothesis and research hypothesis.
c. Can you reject the null hypothesis at $\alpha = .05$? Report your findings in APA format.

13. A nutritionist is trying to determine if a high carbohydrate diet is more or less beneficial than a high protein diet. He recruits a sample of $n = 9$ individuals to go on a carbohydrate diet and another $n = 9$ to go on a protein diet. To reduce the influence of confounding variables, the nutritionist matches each person across the samples according to health prior to starting the diet. He then measures the amount of weight loss over the next 2 weeks. Here are the results:

Carbohydrate	Protein
2	7
1	5
3	5
2	4
1	5
2	4
2	8
3	7
1	9

a. What type of t test should be used?
b. State the null hypothesis and research hypothesis.
c. Can you reject the null hypothesis at $\alpha = .01$? Report your findings in APA format.

14. You are the head researcher for a TV station that has just completed conducting focus groups for a new show that is supposed to appeal to women. The results of an independent-samples t test support this finding as men and women significantly differed on their interest in the show. The mean level of interest (on a scale of 1 to 10) for women was 7.8, whereas the mean level of interest for men was 3.4. The $\sigma_{M_1 - M_2}$ for the study was 1.5 with an $N = 20$. In preparing to present this information to the owner of the TV station, you want to be able to report the possible population gender discrepancy in interest in this show.
a. What would be the point estimate for the mean difference?
b. What would be the 95% confidence interval for this difference?
c. What would be the 99% confidence interval for this difference?

15. The Dr. Johnson from Question 9 is very pleased with the results and wants to tell his fellow professors. To help him ensure that he can feel confident in his results, find the 95% confidence interval.

Answers to Odd-Numbered Computational Exercises

1. $\sigma_{M_1 - M_2} = \sqrt{(7.5)^2 + (4.27)^2} = 8.63$

3. Brown bear: $SS = 1612$; $n = 10$; black bear: $SS = 300.88$; $n = 11$

$$\sigma_{M_1 - M_2} = \sqrt{\left(\frac{1612 + 300.88}{10 + 11 - 2}\right)\left(\frac{1}{10} + \frac{1}{11}\right)} = 4.38$$

5.
   a. Independent variable: monitor size. Dependent variable: hours of computer use.
   b. An independent-samples t test
   c. Null hypothesis: The amount of computer usage by those with a 24-in. monitor does not differ from those with a 15-in. monitor. Alternative hypothesis: The amount of computer usage by those with a 24-in. monitor is greater than for those with a 15-in. monitor.
   d. 24: Mean = 7.78; SS = 25.56; $\sigma^2 = 2.84$
      15: Mean = 2.56; SS = 6.22; $\sigma^2 = 0.69$

   $$\sigma_{M_1 - M_2} = \sqrt{3.19 + 0.78} = 1.88$$

   $$t_{2\text{-samp}} = \frac{(7.78 - 2.56)}{1.99} = 2.78$$

   $df = 16$

   $t_{\text{critical}} < t_{\text{obtained}}$, reject the null hypothesis

   Those with a larger monitor use the computer significantly more than those with a smaller monitor, $t(16) = 2.78$, $p > .05$.

7.
   a. An independent-samples t test
   b. Null hypothesis: There is no relation between wing size and length of flight time. Alternative hypothesis: There is a relation between wing size and length of flight time.

   c. $\sigma_{M_1 - M_2} = \sqrt{\left(\frac{217 + 258}{4 + 8 - 2}\right)\left(\frac{1}{4} + \frac{1}{8}\right)} = 4.22$

   $$t_{2\text{-samp}} = \frac{(8.27 - 10.47)}{4.22} = -0.52$$

   Retain the null hypothesis. It appears there is no relation between wing size and length of flight time, $t(10) = -0.52$, $p > .05$.

9.
   a. Independent variable: class time (8:30 a.m./1:00 p.m.). Dependent variable: approval rating.

b. An independent-samples t test

c. Null hypothesis: There is no difference in Dr. Johnson's approval ratings between his two classes. Alternative hypothesis: There is a difference in Dr. Johnson's approval ratings between his two classes.

d. $\sigma_{M_1 - M_2} = \sqrt{\left(\dfrac{89+95}{16+24-2}\right)\left(\dfrac{1}{16}+\dfrac{1}{24}\right)} = 0.71$

$t_{2\text{-samp}} = \dfrac{3.4 - 5.7}{0.71} = -3.24$

The 1:00 p.m. class likes Dr. Johnson more than the 8:30 a.m. class, $t(38) = -3.24$, $p < .01$.

11.
   a. The independent variable is couple (husband/wife dyad). The dependent variable is marital satisfaction.
   b. A related-samples t test
   c. Null hypothesis: There is no difference between husband's and wife's marital satisfaction. Alternative hypothesis: There is a difference between husband's and wife's martial satisfaction.
   d. Mean difference = −0.5; standard error = 0.87

$t_{2\text{-samp}} = \dfrac{-0.5 - 0}{\sqrt{\dfrac{10 - \dfrac{4}{4}}{4(3)}}} = -0.57$

Retain the null hypothesis. Martial satisfaction is not significantly different between husbands and wives, $t(3) = -0.57$, $p > .05$.

13.
   a. A related-samples t test
   b. Null hypothesis: Carbohydrate and protein diets will equally reduce weight. Alternative hypothesis: Carbohydrate and protein diets will reduce significantly different amounts of weight.
   c. Mean difference = −3.57; standard error = 0.68

$t_{2\text{-samp}} = \dfrac{-3.57 - 0}{\sqrt{\dfrac{185 - \dfrac{1369}{9}}{9(8)}}} = -5.25$

Reject the null hypothesis. Protein diets are more effective than carbohydrate diets at reducing weight, $t(8) = -5.25$, $p < .05$.

15. CI = (3.4 − 5.7) ± (2.04)(0.71)
    95% CI = −3.75 to −0.85

True/False Questions

1. The sampling distribution of the difference between the means consists of the difference between pairs of samples drawn from two populations.

2. The numerator of the two-sample t test formulas compares the obtained difference between the samples with the expected difference between the populations.

3. Participants were divided into two samples such that they were matched on the severity of their depression. These samples are considered independent because they contain different people.

4. The standard error in a related-samples t test must be adjusted to account for the relationship between the participants in the samples.

5. You conduct a study on a new treatment for anxiety. Your results indicate that participants who underwent this new treatment did not have significantly lower anxiety as compared with those who did not undergo the treatment, $t(23) = 1.99, p > .05$. This means the treatment failed in reducing anxiety.

6. There will almost always be a difference between two samples, even if the null hypothesis is true.

7. The standard error of the difference between the means is obtained by pooling the variance of the two samples.

8. A two-sample t test would be appropriate if you were comparing a sample of SAT scores from a single school to the national SAT average.

9. In using a two-sample t test, we must use the estimated population standard deviation when calculating the standard error of the difference between the means.

10. A repeated measures t test uses difference scores rather than raw scores.

11. In an independent-samples t test, the degrees of freedom are $N - 2$, where N is all the participants in the study.

12. A researcher is interested in assessing the effectiveness of a new medication for treating diabetes. She recruits a large sample of diabetics and randomly assigns them to two groups. The first group receives the medication, whereas the second group receives a placebo. The researcher should use a repeated measures t test because all the people in her sample are diabetics.

13. Covariance is a measure of the extent that two scores vary together.

14. In a repeated measures study in the social sciences, it is impossible for the sample sizes in the groups to be unequal.

15. The application of confidence intervals for t tests is very similar to its application for Z tests.

Answers to Odd-Numbered True/False Questions

1. True
3. False

5. False

7. True

9. True

11. True

13. True

15. True

Short-Answer Questions

1. How does the sampling distribution of the mean and the sampling distribution of the difference between the means differ in their creation? How does this influence the meaning of the standard deviation of the sampling distribution of the difference between the means?

2. Why is the formula for a two-sample t test reduced to
$$t_{2\text{-samp}} = \frac{(M_1 - M_2)}{\sigma_{M_1 - M_2}} \text{ from } t_{2\text{-samp}} = \frac{(M_1 - M_2) - (\mu_1 - \mu_2)}{\sigma_{M_1 - M_2}}?$$

3. You are in charge of determining if a new TV show that is targeted at brothers and sisters is enjoyed by both siblings. As a result, you screen the show with $n = 20$ boys and their respective $n = 20$ sisters. In conducting your analysis, should you treat these samples as independent? Why or why not?

4. The formulas for the three different types of two-sample t tests (equal sample sizes, unequal sample sizes, and related samples) all look strikingly different. However, appearances can be deceiving. Are the formulas for the different types of t tests really all that different? Why or why not?

5. You have completed a study that has a $p = .03$. What is your confidence for this study? If this study was a comparison of two sample means, what is the probability that the difference between the means is attributed to sampling error? What is the probability that the difference is attributed to something other than sampling error? Which reason do you think is more likely for the difference between the means?

6. Some researchers are interested in examining the effects of a new medication to treat cancer. To do so, they will conduct a research study in which they will compare the cancer cells in a sample that has received the medication with a sample that has not. What are the null and alternative hypotheses for the study?

7. What two factors influence the size of the standard error of the difference between the means? What two factors influence these two factors?

8. What do each of the bolded items in the following statement tell you? $t(\mathbf{12}) = \mathbf{4.1}$, $\mathbf{p < .05}$.

9. You assess the enjoyment of two vacation spots by asking vacationers to rate their satisfaction on a 1 to 7 scale. You obtain a mean difference between the two samples of 5. However, your two-sample t test does not suggest there is a significant difference in satisfaction between the two spots. You notice that the sample

size of the higher-rated spot is $n = 10$ and the sample size of the lower-rated spot is $n = 25$. What may have prevented your analyses from yielding significant results? What could you do fix this?

10. What is one of the major advantages of using a related-samples t test?

11. Briefly, what does r indicate?

12. How are the formulas for degrees of freedom different for independent-samples t tests and related-samples t tests? What is the cause of this difference?

13. Samantha's elementary school is having a candy-selling competition among the different classes in her grade. Samantha's aunt is a statistician and wants to determine if any of the classes sell significantly more candy. She decides to conduct a related-samples t test because all the classes are in the same school. Is this correct? Why or why not?

14. What is the problem with making a statistical decision a dichotomous decision?

15. How do confidence and a confidence interval differ?

Answers to Odd-Numbered Short-Answer Questions

1. The sampling distribution of the mean is created by hypothetically sampling all possible sample means from a population. The sampling distribution of the difference between the means is created by plotting the differences between all possible samples from the two populations that are being compared. The standard deviation for this distribution represents the average amount of variation expected from the difference in sample means.

3. No, these are not independent. Each participant is matched with a specific member in the other sample.

5. Confidence would be .97. The probability that the difference between the means is attributed to sampling error is .03 or 3%. The probability that the difference between the means is attributed to something else (such as an independent variable) is .97 or 97%. There is a greater chance that the means are differing because of another variable than because of sampling error.

7. The standard error of the difference between the means is influenced by the size of the standard error of the mean for each sample (two samples). The size of each of these standard errors is influenced by the sample size and standard deviation of each sample.

9. The problem may be with the discrepancy in the sample sizes. There may be a restriction of range in the samples or a discrepancy in the variability across the samples. To rectify this situation, you should collect a larger sample.

11. It indicates the strength of the relationship between two variables.

13. This is incorrect. Although all the students are in the same school, each class contains different students, and the students across the different classes were not matched in pairs.

15. Confidence refers to the probability that the difference between two samples is related to the independent variable. Confidence intervals refer to a range of values in which the difference between the two samples is expected to fall if the study were to be done repeatedly.

Multiple-Choice Questions

1. What does pooling the variance indicate?
   a. Combining the variances of two samples to get a single variance
   b. Combining the standard deviations of two samples to get a single variance
   c. Dividing the variance of two groups in two equal proportions
   d. Combining the sums of squares of two samples to get a single sum of squares

2. What would most probably produce a significant result?
   a. A small mean difference and a large $\sigma_{M_1 - M_2}$
   b. A large mean difference and a large $\sigma_{M_1 - M_2}$
   c. A small mean difference and a small $\sigma_{M_1 - M_2}$
   d. A large mean difference and a small $\sigma_{M_1 - M_2}$

3. What is being compared in the numerator of a two-sample t test?
   a. The difference between the two sample means
   b. The difference between the two population means
   c. The difference between the two sample means to the expected difference between the two population means
   d. The difference between the two samples' variabilities

4. You are conducting an independent two-sample t test with an $N = 30$. What are the degrees of freedom for your study?
   a. 30
   b. 28
   c. 15
   d. 13

5. Which of the following results is correct and reported in APA format?
   a. $t(13) = 3.45, p < .05$
   b. $t(13) = 3.45, p > .05$
   c. $t = 3.45, p > .05$
   d. $t = 3.45$

6. For which of the following situations would a repeated measures study be appropriate?
   a. Compare hours of sleep had by adolescents versus senior citizens
   b. Compare weight loss for individuals on a new diet with those on a different diet
   c. Compare salary levels for college graduates with those who did not graduate from college
   d. Compare reaction times before and after taking a pain medication

7. The range of a confidence interval increases when the sample size _____ and the percentage of confidence _____.
   a. Increases, decreases
   b. Increases, increases
   c. Decreases, increases
   d. Decreases, decreases

8. In a study comparing a sample of people taking public transportation in the morning as opposed to the evening, what exactly is being compared in the hypothesis test?
   a. The means of each sample
   b. The means of each population
   c. The standard deviations of each sample
   d. The standard deviations of each population

Part IX Study Guide

9. The mean of the sampling distribution of the difference between the means is always ___.
   a. 1
   b. 2
   c. 3
   d. 0

10. Assuming the null hypothesis is true, this means that there is a low probability that the difference between two samples will be very different from zero. How does this affect the sampling distribution of the difference between the means?
    a. It indicates that the distribution will be positively skewed
    b. It indicates that the distribution will be negatively skewed
    c. It indicates that the distribution will be symmetrical
    d. It indicates that the distribution will be bimodal

11. In a study comparing the distances pigeons can fly, Group 1 has a σ_M^2 of 8. Which value for σ_M^2 would lead to the smallest $\sigma_{M_1 - M_2}$?
    a. 25
    b. 12
    c. 3
    d. 7

12. Which of the following values would allow to you reject the null hypotheses using a two-sample t test with 13 df at $\alpha = .05$, nondirectional test?
    a. 2.10
    b. 2.45
    c. 1.78
    d. 1.97

13. You are conducting a study on a headache relief medicine. You are interested in determining if the medicine provides faster headache relief than aspirin. One sample of $n = 10$ receives the new medication and another sample of the same size receives aspirin. Both samples are asked to report how long it takes for them until they experience pain relief in seconds. You obtain the following data: $M_{new\ medication} = 64$, $\sigma_{est}^2 = 3$; $M_{aspirin} = 81$, $\sigma_{est}^2 = 13$. What is the t value?
    a. 3.24
    b. 4.25
    c. 8.45
    d. 9.23

14. What is the critical value for a two-sample, nondirectional t test with $\alpha = .01$ with 17 df?
    a. 2.01
    b. 2.90
    c. 1.74
    d. 2.57

15. What are the degrees of freedom for a two-sample t test that has a sample with an $n = 10$ and another a sample of $n = 5$?
    a. 15
    b. 13
    c. 10
    d. 5

16. A researcher is examining the influence of a growth hormone on rats. From previous research on this issue, the researchers have the population mean and standard deviation for the growth rate of this species of rat. Which hypothesis test would be most appropriate for this experiment?

a. Normal deviate test
b. One-sample t test
c. Independent-samples t test
d. Related-samples t test

17. A team of researchers is interested in monitoring the influence of a new therapy for children with ADHD. They hope to see a decline in symptoms after 2 weeks of being enrolled in the therapy. Which of the following hypothesis tests would be best suited to determine if the therapy is effective?
    a. Normal deviate test
    b. One-sample t test
    c. Independent-samples t test
    d. Related-samples t test

18. For the study described in Question 17, the researchers observe $n = 12$ children and obtain a mean difference of 4.5 in ADHD symptoms over the 2-week period with an $\sigma_{M_1-M_2} = 1.5$. What is the statistical decision you would make using a nondirectional test?
    a. Reject the null hypothesis at $\alpha = .05$, retain at $\alpha = .01$
    b. Retain the null hypothesis at $\alpha = .05$ and at $\alpha = .01$
    c. Reject the null hypothesis at $\alpha = .05$ and at $\alpha = .01$
    d. Retain the null hypothesis at $\alpha = .05$, reject at $\alpha = .01$

19. In all hypothesis testing, higher levels of α are related to
    a. A higher level of confidence in your decision
    b. A lower level of confidence in your decision
    c. A lower chance of a Type 1 error
    d. a and c

20. A new treatment for depression has been shown to be effective. On a commonly used depression inventory, the mean of the treatment group was 8 and the mean of the control group was 15 with a standard error of the mean difference of 3.2. If $N = 30$ participants were involved in this independent-samples t test, what is the 99% confidence interval (it was a nondirectional test)?
    a. −1.83 to 15.83
    b. 1.83 to 15.83
    c. 4.7 to 10.2
    d. 4.8 to 11.2

Answers to Odd-Numbered Multiple-Choice Questions

1. a
3. c
5. a
7. c
9. d
11. c
13. b
15. b
17. d
19. d

Part X Study Guide

The Multisample Test

Part X Summary

Module 24

- In comparing three groups, you cannot use three separate *t* tests because it unfairly increases the chances of rejecting the null hypothesis. If you were to do a *t* test 100 times using $\alpha = .05$, you could theoretically expect to reject the null hypothesis five times based on chance alone. Therefore, in doing three separate tests to compare three groups using $\alpha = .05$, the chances of falsely rejecting the null hypothesis are raised.
- An *analysis of variance* or *ANOVA* is used to compare three or more groups while maintaining the same level of Type 1 error. ANOVAs rely on the *F* statistic and search for differences in group variances rather than group means.
- There are certain assumptions that must be met to complete an ANOVA. The two primary assumptions are (1) that the populations from which samples are drawn from are normally distributed and (2) that the populations from which the samples are drawn have similar variances. However, these assumptions can be violated to a certain extent when the sizes of the samples are relatively large. An ANOVA is robust to these assumptions.
- ANOVA analyzes variances by splitting the variability of participants into two sections: (1) variability between groups, which is between-treatment variability, and (2) variability within each group, which is within-treatment variability. Between-treatment variability is the extent that the treatment groups differ from each other. This variability is attributed to the independent variable as it is hypothesized to be the main cause for group differences. Within-treatment variability is the extent that members of the same group differ from each other. This variability is considered to be the result of random error because members of the same group receive the same treatment.
- The primary analysis of an ANOVA is done by splitting, or *partitioning*, the total variability into between-treatment and within-treatment variability and then comparing each piece. In partitioning variance, you will need to calculate the means of each group as well as the overall mean (mean of all participants), which is referred to as the *grand mean*. Within-group deviation is the extent that each person differs from his or her own group ($X - M_{group}$). Between-group deviation is the difference of a person's group from the grand mean ($M_{tot} - M_{group}$). Through this process, you can obtain three deviation scores, within-group deviation, between-group deviation, and total deviation (deviation of a score from the grand mean).
- The sum of the squared deviation scores is referred to as the *sum of squares*. When the sum of squares is divided by the total amount of deviation scores, we have obtained a

variance (or the average squared deviation). To find a variance for an ANOVA, you follow a very similar process to that of finding a variance for a sample. However, instead of dividing the sum of squares by n as you would for a sample, you divide the variance by the degrees of freedom. The degrees of freedom are used as they correct the bias that is inherent in using sample statistics to estimate population parameters.

- The variances of an ANOVA are called *mean squares*. The mean square between groups (MS_{bet}) represents the variance attributed primarily to treatment differences, but including some random error. The mean square within groups (MS_{with}) represents the variance attributed to only random error.
- The purpose of an ANOVA is to determine the amount of between-group variance relative to the amount of within-group variance. Thus, the formula for an ANOVA is

$$F = \frac{MS_{bet}}{MS_{with}} \text{ or } F = \frac{\text{Treatment} + \text{Error}}{\text{Error}}$$

- Similar to the t distribution, the F distribution is a family of distributions and is positively skewed. An F ratio can never be negative because variances are always positive. Second, very few F ratios can fall below a score of 1. This would indicate that there is less variability between the groups than within the groups. The majority of situations in which there is no effect will have an equal amount of between- and within-treatment variability. Third, since an F ratio of 1 indicates no effect, it can be assumed that the majority of F ratios will cluster around a score of 1. Thus, $F = 1$ is the modal F ratio. As the treatment effect becomes more pronounced, the ratio will become progressively larger. Therefore, large F ratios occur less frequently and are more statistically significant.

Module 25

- Just as there were different types of t tests, there are different types of ANOVAs. The first ANOVA that is discussed is a one-way ANOVA. *One-way ANOVAs* are used when there is one independent variable.
- Much like all other hypothesis tests, a one-way ANOVA is used to answer the question, "Is the difference among these groups a lot or a little?" As a result, it follows the prototypical formula of all other tests. However, the way in which you compare the differences between your obtained and expected values with the standardized random error is not as apparent. The obtained difference between the groups is the between-group variability. The expected difference between the groups is 0 (as it was with the t tests). The standardized random error in ANOVA refers to the within-group variance, the variability that is the result of random chance.
- The first part of calculating a one-way ANOVA is partitioning the SS into between and within portions. First, the total SS should be found with the following formula:

$$\text{Deviation method: } SS_{tot} = \sum_{1}^{N} (X - M_{tot})^2$$

$$\text{Raw score method: } SS_{tot} = \sum_{1}^{N} X^2 - \sum_{1}^{k} \frac{\left(\sum X_{tot}\right)^2}{N}$$

- Next, the between SS is found with the following formula (k = number of groups):

Deviation method: $SS_{bet} = n_g \sum_{1}^{k} (M_g - M_{tot})^2$

Raw score method: $SS_{bet} = \sum_{1}^{k} \dfrac{\left(\sum X_g\right)^2}{n_g} - \dfrac{\left(\sum X_{tot}\right)^2}{N}$

Finally, the within SS is found with the following formula:

Deviation method: $SS_{with} = \sum_{1}^{k} \sum_{1}^{n} (X - M_g)^2$

Raw score method: $SS_{with} = \sum_{1}^{N} X^2 - \sum_{1}^{k} \dfrac{\left(\sum X_g\right)^2}{n_g}$

There is a great deal of math that is associated with partitioning the variance and so it is important to be organized when you are conducting your ANOVA. Also, there is a "safety net" in place that enables you to check your work; $SS_{tot} = SS_{bet} + SS_{with}$. You should always add up your SSs after conducting an ANOVA to ensure that no mistakes were made.

- After the variance has been partitioned, the next step is to find the appropriate degrees of freedom. The formulas for the degrees of freedom are as follows:

  df_{bet} = No. of groups − 1 = $k - 1$
  df_{with} = No. of subjects − No. of groups = $N - K$
  df_{tot} = No. of subjects − 1 = $N - 1$

- After obtaining the degrees of freedom and the SS, we can now find the mean square by dividing each SS by its corresponding degrees of freedom.

$$MS_{bet} = \dfrac{SS_{bet}}{df_{bet}}$$

$$MS_{with} = \dfrac{SS_{with}}{df_{with}}$$

- The final step in this process is to calculate the F value, which is done by dividing the MS_{bet} by the MS_{with}.
- Statistical significance for an F value is similar and different to that of a t test. It is similar in that it continues to compare deviations from an expected result with random error. However, it differs in that it cannot use the normal curve to determine if this difference is a lot or a little. As a result, a different table is used for F ratios, which can be found in Appendix D. Remember, a significant F ratio will be one that is very different from a value of 1. The table requires the use of the two degrees of freedom that were used to calculate the F ratio, the df_{bet} and the df_{with}. The obtained F ratio is then compared with the critical F ratio. If the F ratio from the ANOVA (obtained F) is larger than the one from the table (critical F), the null hypothesis is rejected. Otherwise, the null hypothesis is retained.

- The results of an ANOVA are presented in APA format in the following fashion: $F(df_{bet}, df_{with}) = F_{obtained}, p < .05$.
- All the calculations and steps involved in an ANOVA can be difficult to organize and present in a neat format. An efficient organization of all the pertinent information can be presented in an *ANOVA summary table*, which appears as follows:

Source	SS	df	MS	F
Between treatments	SS_{bet}	df_{bet}	SS_{bet}/df_{bet}	MS_{bet}/MS_{with}
Within treatments	SS_{with}	df_{with}	SS_{with}/df_{with}	
Total	SS_{tot}	df_{tot}		

Learning Objectives

Module 24

- Know the assumptions underlying analysis of variance
- Distinguish between conditions under which it is appropriate to conduct a *t* test versus an ANOVA
- Understand the relationship of within, between, and total variances
- Understand how the possible values of *F* follow from the distribution's shape
- Understand how changes in the within- or between-group variances influence the *F* statistic

Module 25

- Understand the similar logic underlying various test statistics
- Determine degrees of freedom
- Caluclate a one-way ANOVA
- Use a table to interpret calculated *F*
- Display the results in an ANOVA summary table
- Report results in APA format

Computational Exercises

1. Partition the variability for the following scores (find the *SS* between, within, and total):

1	2	3
9	5	8
1	10	6
5	3	9
3	8	3
7	2	1

2. Find the *MS* for the values in Question 1. Without finding the *F* ratio, do you suspect it will be a significant value? Why or why not?

3. Partition the variability for the following scores (find the *SS* between, within, and total).

1	2	3
3	10	8
2	8	10
6	7	8
5	3	10
1	7	9
2	2	9

4. You are conducting research on a new therapy for social anxiety by comparing those who receive the therapy with those who do not. However, you want to avoid the possibility that people are improving just because they are interacting with another person. To address this confound, you add a group of participants who will meet with a therapist on a schedule similar to that of the social anxiety group, but will not receive therapy. This group will be referred to as the attention-control group. Here are the data you obtain from a measure of social anxiety with a scale of 0 to 20:

Therapy	*Control*	*Attention-Control*
2	19	14
7	13	15
5	20	13
4	14	12
2	12	9

   a. What are the null and alternative hypotheses for this study?
   b. Make a statistical decision at the .05 level to determine if there is a significant difference in anxiety among the therapies.
   c. Write the results in APA format.

5. A local police department has divided the city into three sections. The police chief wants to determine if officers are biased in the number of parking tickets they give out in each section. Here are the data for the number of parking tickets given out by the 6 officers who work in each section of the city:

Area 1	*Area 2*	*Area 3*
8	3	1
4	7	2
6	0	7
8	2	6
6	7	5
4	5	0

   a. What are the null and alternative hypotheses for this study?
   b. Make a statistical decision at the .05 level to determine if there is a significant difference in the number of tickets issued among the sections.
   c. Write the results in APA format.

6. A music research group is interested in determining which of the three most popular music acts has the most regular concert attendance. They interview 6 fans of

each music act and ask them how many concerts they have attended. Here are the data they obtain:

Act 1	Act 2	Act 3
4	4	0
3	1	1
5	2	0
2	1	1
3	4	0
5	3	2

    a. What are the null and alternative hypotheses for this study?
    b. Make a statistical decision at the .01 level to determine if there is a significant difference in attendance among the acts.
    c. Write the results in APA format.

7. Three different statistics professors are asked by their department to determine if they are teaching reliably (this would be determined by there being no difference in the performance of the students across the classes). The statistics professors decide to take a random sample of grades from each of their classes and conduct an ANOVA to determine if there are any differences among the classes. Here are the data they obtain:

Class 1	Class 2	Class 3
85	92	76
93	80	83
85	93	77
98	75	94
80	92	100

    a. What are the null and alternative hypotheses for this study?
    b. Make a statistical decision at the .01 level to determine if there is a significant difference in performance among the classes.
    c. Write the results in APA format.

8. Three nail salons are competing to determine which salon can provide the longest lasting manicure. They compare the length of time it takes for $n = 5$ customers to require a new manicure. Here are the data they obtain:

Salon 1	Salon 2	Salon 3
27	23	17
24	27	21
20	21	24
26	12	25
15	10	16

    a. What are the null and alternative hypotheses for this study?
    b. Make a statistical decision at the .05 level to determine if there is a significant difference in manicure endurance among the salons.
    c. Write the results in APA format.

9. A construction company is using a new adhesive in their new building. They want to determine on which surface the adhesive works best: wood, steel, or concrete.

To determine this, they use the adhesive on each surface four different times and determine how much pressure (in pounds) can be applied before the adhesive fails. Here are the data they obtain:

Wood	Steel	Concrete
97	141	94
88	132	101
98	134	103
84	129	91

   a. What are the null and alternative hypotheses for this study?
   b. Make a statistical decision at the .05 level to determine if there is a significant difference in adhesive endurance among the surfaces.
   c. Write the results in APA format.

10. A pharmaceutical company is testing the liver toxicity of a new drug for alcoholism. They obtain a sample of $N = 21$ alcoholics and randomly assign them to one of three groups: treatment, control, and placebo. Participants are asked to take the medication for 4 weeks, after which the effectiveness of their liver enzymes is assessed on a 0 to 100 scale. Here are the data they obtain:

Drug	Control	Placebo
92	26	15
93	33	53
89	53	55
95	45	22
36	45	84
79	71	42
89	24	90

   a. What are the null and alternative hypotheses for this study?
   b. Make a statistical decision at the .05 level to determine if there is a significant difference in liver toxicity among the drugs.
   c. Write the results in APA format.

11. A cognitive psychologist is interested in studying the influence of stress on reaction time. She obtains a sample of $N = 30$ participants and randomly assigns them to one of three groups: calm, stressed, and highly stressed. The stressed groups are asked to watch a short film clip that will induce anxiety. The researcher obtains the following information from her participants:

	Calm	Stressed	Highly Stressed
Mean	230	450	497
SS	5,478	5,347	5,317

   a. The $SS_{tot} = 18,542$. Create a source table that will address the primary hypothesis of the researcher. What are the null and alternative hypotheses for this study?
   b. Make a statistical decision at the .05 level to determine if there is a significant difference in reaction time among the stress conditions.
   c. Write the results in APA format.

12. Pet Lover's magazine is conducting its annual poll to determine how much affection different pet owners have for their pets. They send out a survey asking pet owners

to rate their love for their animals on a 0 to 50 scale and have 5 different categories of pet: dog, cat, fish, turtle, and rabbit. They receive a grand total of 100 responses, with an equal number of responders falling in each pet category. Here are the data they obtained from their survey:

	Dog	Cat	Fish	Turtle	Rabbit
Mean	39	34	30	24	28
SS	210	324	245	207	547

The $SS_{tot} = 2{,}000$. Create a source table that will address whether there are differences between the love different pet owners have for their pets.

  a. What are the null and alternative hypotheses for this study?
  b. Make a statistical decision at the .05 level to determine if there is a significant difference in owner affection among the pets.
  c. Write the results in APA format.

13. A prominent ice cream company is trying to discover which ice cream flavor is most preferred by their customers. They have narrowed down their choices to mint chocolate chip, butter pecan, pistachio, and cookie dough. They decide to investigate daily sales of each flavor over the course of 2 weeks ($n = 14$ days) to determine which flavor is most preferred. Here are the data they have obtained:

	Mint Chocolate Chip	Butter Pecan	Pistachio	Cookie Dough
Mean	23	24	31	37
SS	78	84	81	92

The $SS_{tot} = 378$. Create a source table that will address the question of whether there is a significant difference among the preferences for the flavors.

  a. What are the null and alternative hypotheses for this study?
  b. Make a statistical decision at the .05 level to determine if there is a significant difference in preference among the flavors.
  c. Write the results in APA format.

14. A school chancellor has very limited funding for the schools in her district. She is able to narrow down the three schools with the most need, but is unable to determine which school would benefit from the most funding. To help with her decision, she consults the annual reading scores of each school over the past 7 years to determine which school would benefit the most. Here are the data she obtains:

	School 1	School 2	School 3
Mean	63	61	58
SS	201	213	278

The $SS_{tot} = 954$. Create a source table that will help the chancellor decide who receives the funding.

  a. What are the null and alternative hypotheses for this study?
  b. Make a statistical decision at the .05 level to determine if there is a significant difference in need among the schools.
  c. Write the results in APA format.

Answers to Odd-Numbered Computational Exercises

1. $SS_{bet} = 0.93$; $SS_{with} = 130.4$; $SS_{tot} = 131.33$

3. $SS_{bet} = 102.11$; $SS_{with} = 69.67$; $SS_{tot} = 171.78$

5.
   a. Null hypothesis: There is no difference in the number of tickets distributed by the police in the three areas. Alternative hypothesis: There is a difference in the number of tickets distributed by the police in the three areas.
   b. Means: 1 = 6; 2 = 4; 3 = 3.5
      SS: $SS_{bet} = 21$; $SS_{with} = 97.5$; $SS_{tot} = 118.5$
      df: $df_{bet} = 2$; $df_{with} = 15$; $df_{tot} = 17$
      MS: $MS_{bet} = 10.5$; $MS_{with} = 6.5$
      $F = 1.62$
      Retain the null hypothesis.
   c. There was no significant difference among the tickets distributed across the three areas of the city, $F(2, 15) = 1.62, p > .05$.

7.
   a. Null hypothesis: There is no difference in grades of the students across the three classes. Alternative hypothesis: There is a difference in grades of the students across the three classes.
   b. Means: 1 = 88.2; 2 = 86.4; 3 = 86
      SS: $SS_{bet} = 13.73$; $SS_{with} = 934$; $SS_{tot} = 947.73$
      df: $df_{bet} = 2$; $df_{with} = 12$; $df_{tot} = 14$
      MS: $MS_{bet} = 6.87$; $MS_{with} = 77.83$
      $F = 0.09$
      Retain the null hypothesis.
   c. There is no significant difference among the performance of students in the three different classes, $F(2, 14) = 0.09, p > .05$.

9.
   a. Null hypothesis: There is no difference in the strength of the adhesive on different surfaces. Alternative hypothesis: There is a difference in the strength of the adhesive on different surfaces.
   b. Means: wood = 91.75; steel = 134; concrete = 97.25
      SS: $SS_{bet} = 4221.17$; $SS_{with} = 315.5$; $SS_{tot} = 947.72$
      df: $df_{bet} = 2$; $df_{with} = 9$; $df_{total} = 11$
      MS: $MS_{bet} = 2110.58$; $MS_{with} = 35.06$
      $F = 60.21$
      Reject the null hypothesis.
   c. There is a significant difference in the strength of the adhesive on different surfaces, $F(2, 9) = 60.21, p < .05$.

11.
   a. Null hypothesis: Stress level has no impact on reaction time. Alternative hypothesis: Stress level has an impact on reaction time.
      Reject the null hypothesis.
   b.

	SS	df	MS	F
Between treatments	2400.00	2.00	1200.00	2.01
Within treatments	16142.00	27.00	597.85	
Total	18542.00	29.00		

   Retain the null hypothesis.
   c. Stress level does not significantly affect reaction time, $F(2, 27) = 2.01$, $p > .05$.

13.
   a. Null hypothesis: There is no difference among the preferences for different types of ice cream. Alternative hypothesis: There is a difference among the preferences for different types of ice cream.
   b.

	SS	df	MS	F
Between treatments	43.00	3.00	14.33	0.43
Within treatments	335.00	10.00	33.50	
Total	378.00	13.00		

   Retain the null hypothesis.
   c. There is no difference amongst the preferences for the different types of ice cream, $F(3, 10) = 0.43$, $p > .05$.

True/False Questions

1. ANOVA uses a completely different method of inferential logic as compared with a *t* test or normal deviate test.

2. ANOVA enables you to compare more than two groups while maintaining the same level of Type 1 error.

3. ANOVAs compare group mean differences.

4. ANOVA is robust to the assumption of normality with larger sample sizes.

5. When variance is partitioned in an ANOVA, it is divided into two equal parts.

6. The grand mean is the mean of all scores in the analysis.

7. Between-group variability contains only variability attributed to the effect.

8. Within-group variability contains only variability due to error.

9. The mean square has a very similar interpretation to a variance.

10. An F less than 1 has a very high chance of being significant.

11. A one-way ANOVA compares differences in one group.

12. The F distribution appears very similar to the z and t distributions.

13. In a perfect situation, all individuals in the same group would have the exact same score.

14. Three groups have different group means, but there is a great amount of variability in each group. This situation will definitely yield a significant F ratio.

15. A strong treatment effect will be related to a large numerator in computing the F ratio.

Answers to Odd-Numbered True/False Questions

1. False
3. False
5. False
7. False
9. True
11. False
13. True
15. True

Short-Answer Questions

1. Mr. Jones wants to compare the overall GPA of his four children. Why can't he use multiple t tests?

2. What are the assumptions that must be met to use an ANOVA?

3. What does it mean that ANOVA is robust to its assumptions?

4. What is within-treatment variability attributed to? How does this differ from between-treatment variability?

5. A t test compares mean differences. What does an ANOVA compare?

6. Demonstrate how a single person's score can be partitioned into between-group and within-group variance.

7. Why do you divide the SS by the df in an ANOVA instead of n?

8. Explain why the F distribution appears as positively skewed.

9. What would cause an F ratio to be statistically significant?

10. Can an F ratio test be directional? Why or why not?

11. What is a mean square? Is it the same or different from a variance?

12. The numerator in an *F* ratio contains random error. Where did this random error come from?
13. In writing the results of an ANOVA in APA format, what degrees of freedom values must be reported?
14. What are the primary steps in conducting an ANOVA?
15. In conducting a research study, what should the researchers attempt to do in recruiting their sample in an attempt to increase the *F* ratio?

Answers to Odd-Numbered Short-Answer Questions

1. He would be unfairly increasing the chances of obtaining a significant result. In other words, he is inflating the chances of making a Type 1 error.
3. It means that with large enough samples, the assumptions that underlie the use of ANOVA can be violated and the results will still be valid.
5. An ANOVA compares variances. The variability between treatment groups is compared with the variability within each treatment group.
7. The *SS* is divided by the *df* because the variances in an ANOVA are estimated population variances. Since the *df* subtracts at least one from the denominator, it enables the resulting mean square to be a population estimate.
9. A significant *F* ratio is the result of a large amount of between-group variability (large group differences) and a minimal amount of within-group variability (very few differences among those in the same groups).
11. A mean square is the average of the squared deviations. It is essentially the same as a variance. However, it is important to note that a mean square is an estimate of the population variance.
13. The degrees of freedom for the numerator (between) and those for the denominator (within).
15. There are a number of things the researcher can do. However, in the context of this chapter, the best thing a researcher could do is try to ensure that all the participants in the sample are very similar. This will minimize the effect of any differences among the individuals and reduce the within-treatment variability.

Multiple-Choice Questions

1. What is the *F* ratio of a treatment that had no effect?
    a. Infinity
    b. 0
    c. About 100
    d. About 1

2. What can an *F* ratio never be?
    a. 1
    b. 1,000
    c. 0.54
    d. −2

3. What other statistical term is the mean square most similar to?
   a. Mean
   b. Standard deviation
   c. Variance
   d. Sum of squares

4. In an ANOVA, a statistically significant F ratio contains a lot of _____ variability as compared with _____ variability.
   a. Between-treatment, within-treatment
   b. Between-subject, within-subject
   c. Within-treatment, between-treatment
   d. Within-subject, between-subject

5. Between-group variability is a measure of
   a. The deviation of each score from the grand mean
   b. The deviation of each score from its group mean
   c. The deviation of each group mean from the grand mean
   d. Variability attributed to random error

6. In a data set containing three groups, the total SS is 431, and the between SS is 123. What is the within SS?
   a. 308
   b. 554
   c. 41
   d. 123

7. For a data set containing three groups, $SS_{tot} = 8{,}792$; $SS_{Group\ 1} = 427$; $SS_{Group\ 2} = 574$; $SS_{Group\ 3} = 177$. What is the SS_{bet}?
   a. 1,178
   b. 7,614
   c. 5,148
   d. 3,176

8. A research study has a total of $N = 400$ participants across 8 groups. If there are an equal number of participants in each group, what are the degrees of freedom for this study?
   a. $df_{bet} = 7$; $df_{with} = 400$; $df_{tot} = 399$
   b. $df_{bet} = 8$; $df_{with} = 392$; $df_{tot} = 399$
   c. $df_{bet} = 24$; $df_{with} = 247$; $df_{tot} = 399$
   d. $df_{bet} = 7$; $df_{with} = 392$; $df_{tot} = 399$

9. A research study containing three groups has an $n = 14$ participants per group. What are the degrees of freedom for this study?
   a. $df_{bet} = 2$; $df_{with} = 39$; $df_{tot} = 42$
   b. $df_{bet} = 2$; $df_{with} = 39$; $df_{tot} = 41$
   c. $df_{bet} = 39$; $df_{with} = 2$; $df_{tot} = 42$
   d. $df_{bet} = 42$; $df_{with} = 39$; $df_{tot} = 2$

10. What information does an ANOVA (and just an ANOVA) provide you with?
    a. Whether or not there is a significant difference between groups at a set α level
    b. Which group had the significantly highest mean
    c. Which group had the significantly lowest mean
    d. The amount that two group means must differ in order to be considered significantly different

11. Which value of the mean square between is most indicative of no group differences?
    a. 1
    b. −1
    c. 0
    d. Infinity

12. A research study has $N = 12$ participants in 3 groups (an equal number of participants per group). What is the F critical value at $\alpha = .01$?
    a. 3.88
    b. 6.93
    c. 3.49
    d. 5.95

13. The research study in Question 12 yields an F value of 6.32. Which of the following correctly shows the result in APA format?
    a. $F(2, 12) = 6.32, p < .01$
    b. $F(3, 12) = 6.32, p > .01$
    c. $F(2, 9) = 6.32, p > .01$
    d. $F(2, 9) = 6.32, p < .01$

14. A study of the effectiveness of a new material for contact lenses has $n = 10$ people per group in four different groups. The results indicate that the $SS_{tot} = 2,147$ and $SS_{bet} = 354$. What is the MS_{with}?
    a. 17,193
    b. 36.21
    c. 49.81
    d. 88.5

15. Using the information from Question 14, what is the F ratio?
    a. 88.5
    b. 49.81
    c. 2.37
    d. 5.23

16. Using the information from the previous question, what conclusion would you draw for $\alpha = .05$?
    a. $F(4, 10) = 2.37, p > .05$
    b. $F(4, 10) = 5.23, p > .05$
    c. $F(3, 39) = 2.37, p > .05$
    d. $F(3, 39) = 5.23, p > .05$

17. Below is the source table for a study on depression comparing two different treatments with a placebo group. What is the F ratio?

	SS	df	MS	F
Between treatments	782.00			
Within treatments		34		
Total	2147.00			

    a. 9.74
    b. 7.25
    c. 8.41
    d. 6.23

18. Using the information from Question 17, how many participants were involved in the study?
    a. 36
    b. 34
    c. 39
    d. 37

19. A study was done comparing how well students performed on a final exam in three different rooms: the original room in which they had the class, a room different from the one in which they had the class, or in a different room but thinking about the original room. In all, 60 students were involved with an equal number of students in each group. Here are the SS values for each group: same = 475; different = 427; think = 412. Using the following source table, what is the F value of this study?

	SS	df	MS	F
Between treatments	1,159	2	579.5	
Within treatments	1,314	57	23.05	
Total	2,473	59		

   a. 32.21
   b. 14.21
   c. 23.05
   d. 25.14

20. Which would be the correct way of writing the obtained result in APA format from the previous question?
    a. $F(2, 59) = 25.14, p < .05$
    b. $F(2, 57) = 25.14, p < .01$
    c. $F(3, 59) = 23.05, p > .05$
    d. $F(2, 59) = 25.14, p > .05$

Answers to Odd-Numbered Multiple-Choice Questions

1. d
3. c
5. c
7. b
9. b
11. c
13. c
15. c
17. a
19. d

Part XI Study Guide

Post Hoc Tests

Part XI Summary

Module 26

- An ANOVA (*F* test) tells you that there is a significant difference among the groups that were compared. However, it does not specify which groups are significantly different. To determine which groups are significantly different from one another, you must conduct a *post hoc* test. If you do not have a significant *F* ratio, then a post hoc test is unnecessary.
- There are many different types of post hoc tests, but all have the same function; determining significant differences among the groups that were compared in an ANOVA. Unlike multiple *t* tests, post hoc tests do not increase the chance of making a Type 1 error. This is because post hoc tests use a fixed error rate. In other words, a post hoc test "spreads the .05 around" the multiple comparisons to ensure that the chances of making a Type 1 error remain stable.
- One post hoc test is called the Tukey HSD, or *honestly significant difference*. If the difference between two group means exceeds the value obtained by the HSD, then they are considered significantly different. If the difference between groups is less than the value of the HSD, then they are not significantly different. For example, if the HSD = 3, then means of 6 and 4 would not be significantly different, because the difference between these means is 6 – 4 = 2, which is less than the HSD. However, means of 6 and 2 would be significantly different because their difference of 6 – 2 = 4 exceeds the value of the HSD. The formula for the HSD is as follows:

$$\text{HSD} = q\sqrt{\frac{MS_{\text{with}}}{n_g}}$$

- In the formula above, a new statistic is presented, the *q* statistic, which refers to the *studentized ranged statistic*. *q* is a correction factor that helps the post hoc test maintain a steady level of Type 1 error. To find *q*, refer to the table in Appendix E. *q* is found by using *k* (the number of groups) and the df_{with}.

Module 27

- The Tukey HSD is just one of the many types of post hoc tests. A second type of post hoc test is called the *Scheffé test*. The Scheffé test creates a series of two-group ANOVAs in which each treatment is compared with all other treatments. The formula for the Scheffé test is as follows:

$$F_{\text{Factor 1 and Factor 2}} = \frac{MS_{\text{between Factor 1 and Factor 2}}}{MS_{\text{within}}}$$

- To find the numerator for a Scheffé test, you must first recalculate the MS for each pair. This is done by using a slight variation of the original formula for the between-treatments SS. Rather than using the SS of all three groups subtracted from the total, the SS for each individual group is added and subtracted from the combined SS for both groups. Once this new SS has been obtained, it is divided by the df from the original ANOVA to obtain the MS for the two groups. By using the original df, the Type 1 error rate is maintained. The final step is dividing this new MS by the MS_{within} from the original ANOVA.
- The final step in a Scheffé test is determining if there is a significant group difference. This is done in a similar manner to the original F test. Find the F critical by using the F table. If the F ratio from the Scheffé test exceeds the F critical value, then the means are significantly different.

Learning Objectives

Module 26

- Distinguish between the information provided by an F test and a Tukey HSD
- Know when it is appropriate to calculate a Tukey HSD
- Calculate a Tukey HSD
- Construct a matrix of group mean differences
- Report the results in APA format

Module 27

- Distinguish between the information provided by an F test and a Scheffé test
- Know when it is appropriate to calculate a Scheffé test
- Calculate a Scheffé test
- Report the results in APA format

Computational Exercises

1. Using the tables below, determine which groups are significantly different at $\alpha = .05$.

	SS	df	MS	F
Between	214.00	3.00	71.33	10.90
Within	314.00	48.00	6.54	
Total	528.00	51.00		

	1	2	3	4
Mean	32.00	24.00	67.00	54
N	5	5	5	5

2. Using the tables below, determine which groups are significantly different at $\alpha = .05$.

	SS	df	MS	F
Between	327.00	2.00	163.50	9.66
Within	457.00	27.00	16.93	
Total	784.00	29.00		

	1	2	3
Mean	20.00	34	45
N	10	10	10

3. Using the tables below, determine which groups are significantly different at $\alpha = .05$.

	SS	df	MS	F
Between	32.00	2.00	16.00	12.16
Within	75.00	87.00	1.32	
Total	107.00	89.00		

	1	2	3
Mean	20.00	34	45
N	30	30	30

4. Refer back to Question 12 (type of pet) in the computational exercises of the previous module. Which group was significantly different at $\alpha = .05$?

5. Using the following tables, compute a Tukey HSD. There were an equal number of participants per group at $\alpha = .05$.

	1	2	3	4
Mean	47.00	42.00	34.00	21.00

	SS	df	MS	F
Between	578.00	3.00	192.67	6.24
Within	987.00	32.00	30.84	
Total	1565.00	35.00		

6. Using the following tables, compute the post hoc test as much as you can with the information provided, at α = .05.

	1	2	3
Mean	47.00	42.00	34.00
n	12	12	12

	SS	df	MS	F
Between	478.00	2.00	239.00	5.43
Within	1453.00	33.00	44.03	
Total	1931.00	35.00		

7. Here is the data from Question 11 (stress level) in the computational exercises of the previous module. Determine where the significant differences fall at α = .05.

	Calm	Stressed	Highly Stressed
Mean	230	450	497
SS	5478	5347	5317

	SS	df	MS	F
Between treatments	2400.00	2.00	1200.00	2.01
Within treatments	16142.00	27.00	597.85	
Total	18542.00	29.00		

8. Here is the source table from Question 10 (alcoholism) in the computational exercises of the previous module. Conduct a Turkey HSD test to determine which groups are significantly different at α = .05.

	SS	df	MS	F
Between treatments	5962.67	2	2981.33	5.9
Within treatments	9098.29	18	505.46	
Total	15060.95	20		

	Drug	Control	Placebo
Mean	81.86	42.43	51.57
n	7	7	7

9. Using the raw data from Question 10 in the computational exercises of the previous module (which was used in the previous question), calculate the significant differences using a Scheffé test at α = .05. Do the results differ based on which post hoc test was used?

Drug	Control	Placebo
92	26	15
93	33	53
89	53	55
95	45	22
36	45	84
79	71	42
89	24	90

10. Using the following data, determine which groups are significantly different using a Scheffé test at α = .05.

Group 1	Group 2	Group 3
67	72	93
50	49	76
68	72	86
26	40	71
74	35	95
67	72	93
50	49	76

	SS	df	MS	F
Between	2812.93	2.00	1406.47	5.29
Within	3188.00	12.00	265.67	
Total	6000.93	14.00		

11. Using the following data, determine which groups are significantly different using a Scheffé test at α = .05.

Group 1	Group 2	Group 3
50	43	69
55	44	53
58	40	62
57	58	70
45	53	68
45	55	71

	SS	df	MS	F
Between	954.33	2.00	477.17	10.47
Within	683.67	15.00	45.58	
Total	1638.00	17.00		

12. Using the following data, determine which groups are significantly different using a Scheffé test at α = .05.

Group 1	Group 2	Group 3
2	14	11
3	2	23
7	6	24
10	0	20
5	0	19
14	9	17
5	10	11

	SS	df	MS	F
Between	634.38	2.00	317.19	12.76
Within	447.43	18.00	24.86	
Total	1081.81	20.00		

13. Using the following data, determine which groups are significantly different using a Scheffé test at α = .05.

Group 1	Group 2	Group 3
19	19	7
26	26	7
23	28	14
11	26	8
27	24	15
10	15	6
21	26	14
15	21	10
23	28	10
12	16	8

	SS	df	MS	F
Between	880.27	2.00	440.13	17.79
Within	667.90	27.00	24.74	
Total	1548.17	29.00		

14. Using the following data, determine which groups are significantly different using a Scheffé test at α = .05.

Group 1	Group 2	Group 3	Group 4
13	18	25	26
14	11	35	22

Group 1	Group 2	Group 3	Group 4
11	10	36	25
10	15	34	28
12	11	36	24
20	20	30	21
16	17	33	22
20	18	28	18
18	14	34	27
19	16	32	16

	SS	df	MS	F
Between	1953.02	3.00	651.01	33.21
Within	823.44	42.00	19.61	
Total	2776.46	45.00		

Answers to Odd-Numbered Computational Exercises

1. $3.79\sqrt{\frac{6.54}{5}} = 4.33$

   All the groups are significantly different from one another.

3. $3.74\sqrt{\frac{1.32}{30}} = .71$

   All the groups are significantly different from one another.

5. $3.85\sqrt{\frac{31.84}{9}} = 7.24$

   Groups 1 and 2 are significantly different from Groups 3 and 4. Group 3 and Group 4 are significantly different as well.

7. $3.9\sqrt{\frac{597.85}{10}} = 30.16$

   All the groups are significantly different from each other.

9.

	SS	MS	F
Drug and control =	5441.143	2720.571	5.382367
Drug and placebo =	3210.286	1605.143	3.175608
Placebo and control =	292.5714	146.2857	0.289411

The only difference is between the drug and control group. The results are identical to the HSD test.

11.

	SS	MS	F
1 and 2	24.08	12.04	0.26
1 and 3	574.08	287.04	6.30
2 and 3	833.33	416.67	9.14

Groups 1 and 2 are significantly different from Group 3.

13.

	SS	MS	F
1 and 2	88.20	44.10	1.80
1 and 3	387.20	193.60	7.91
2 and 3	845.00	422.50	17.27

Groups 1 and 2 are significantly different from Group 3.

True/False Questions

1. A post hoc test is exactly like doing multiple t tests.
2. Although there are many post hoc tests, they all provide the same results.
3. The Tukey HSD test can only be used when the groups have an equal number of participants.
4. The Scheffé test is more conservative in its allocation of Type 1 error than the Tukey HSD test.
5. When calculating the $MS_{between}$ for a Scheffé test, you should use the $df_{between}$ from the original ANOVA.
6. The q statistic is used to maintain the level of Type 1 error in a Tukey HSD.
7. q is found by using the $df_{between}$ and the df_{within}.
8. Means with a difference less than that obtained by the HSD are significantly different.
9. In a Tukey HSD test, the denominator is the number of participants in the study.
10. A Scheffé test determines significant differences by comparing all the possible pairs of means.
11. The MS_{within} for a Scheffé test is obtained from the original ANOVA.
12. The Scheffé test requires a different $MS_{between}$ for each comparison.
13. The $SS_{between}$ in a Scheffé test is found by dividing the total $SS_{between}$ from the original ANOVA.
14. You are less likely to make a Type 1 error with a Tukey HSD than with a Scheffé test.
15. A Scheffé test provides you with one F ratio for determining significant differences.

Answers to Odd-Numbered True/False Questions

1. False
3. True
5. True
7. False
9. False
11. True
13. False
15. False

Short-Answer Questions

1. Why are post hoc tests necessary?
2. Chester has just completed his first ANOVA with the result of $F(2, 10) = 1.23, p < .05$. Should he do a post hoc test? Why or why not?
3. Why is an ANOVA/post hoc test preferable to conducting numerous t tests?
4. What is the purpose of the q statistic?
5. What pieces of information are used to determine the q statistic?
6. What is an HSD?
7. How do Scheffé and Tukey HSD tests differ?
8. Which post hoc test is considered more conservative? Why?
9. Using the formula, how does a Scheffé test maintain the Type 1 error rate?
10. What is the difference between a Scheffé F test and an ANOVA F test?
11. What are the steps of conducting a Scheffé test?
12. If you were to obtain a significant F ratio for a Scheffé test, how would you know which groups were significantly different? How does this differ from a Tukey HSD test?
13. A dry cleaner was comparing the effectiveness of three different detergents on removing a stain from a shirt. A significant F ratio was found and the detergents' post hoc test revealed an HSD of 3.2. Using the means in the following table, interpret the findings.

	Detergent 1	Detergent 2	Detergent 3
Mean	7.20	9.00	11.30

14. A CEO is comparing the scores on an aptitude test of three different departments in her company. Unfortunately, she will have to let the lowest scoring department go at the end of the financial year. If she found a significant F ratio, indicating that one group's performance was significantly different from that of another group,

which group should she fire and what should she tell them? (Use the following table of results; there were 5 people in each department.)

	SS	MS	F
1 and 2	1960.82	653.61	0.83
1 and 3	6181.35	2060.45	2.61
2 and 3	9652.02	3217.34	4.08

	Department 1	Department 2	Department 3
Mean	89	91	86

15. The following means represent the amount of hazardous emissions put out by the cars of different car manufacturers. If there are an equal number of cars made by each manufacturer and the HSD = 5.2, which cars are putting out the most hazardous emissions?

	Manufacturer 1	Manufacturer 2	Manufacturer 3	Manufacturer 4	Manufacturer 5
Mean	12.00	13.00	10.00	8.00	21

Answers to Odd-Numbered Short-Answer Questions

1. They determine which groups are significantly different in an ANOVA. This information is not provided from the original F test.

3. ANOVA and post hoc tests maintain a constant level of Type 1 error whereas multiple t tests will raise the chances of making a Type 1 error.

5. The number of groups in the variable and the df_{within}.

7. The Scheffé test is comparing the amount of variability between two groups (the recalculated $SS_{between}$) to the total amount of error variability in the problem. Significance is then determined with an F ratio. An HSD determines the amount by which two means must differ to be considered significantly different. This is based on a portion of the error variability and a correction factor (q).

9. A Scheffé test maintains the Type 1 error rate by using the MS_{within} in calculating the F ratio.

11. (1) Determine the $SS_{between}$ for the two groups being compared. (2) Determine the MS for the two groups by dividing by the original $df_{between}$. (3) Divide the $MS_{between}$ for the two groups by the MS_{within} to obtain the F ratio. (4) Compare the obtained F ratio to a critical F ratio obtained by using the original $df_{between}$ and df_{within}.

13. Detergent 3 appears to be the most effective detergent because it is significantly better than Detergent 1. Detergent 2 is comparable to both Detergents 1 and 3, which indicates that Detergent 3 may be superior, considering that it has the highest mean. There may have been evidence of a Type 2 error in the difference between Detergents 2 and 3.

15. It appears that Manufacturer 5 is making the car with the most hazardous emission output. The rest of the manufacturers are not significantly different from one another.

Multiple-Choice Questions

1. How many groups are compared per F ratio in a Scheffé test?
   a. 2
   b. 3
   c. 4
   d. As many as there are in the ANOVA

2. Five groups are compared in a single ANOVA. How many F ratios will need to be computed when doing a Scheffé post hoc test?
   a. 5
   b. 7
   c. 8
   d. 10

3. Using the following mean table and an HSD = 4, how many groups are significantly different from the first?

	Group 1	Group 2	Group 3	Group 4
Mean	4	7	12	13

   a. 1
   b. 2
   c. 3
   d. 4

4. Using the following ANOVA source table, compute the Tukey HSD at $\alpha = .05$.

	SS	df	MS	F
Between	25.00	3.00	8.33	5.21
Within	32.00	20.00	1.60	
Total	57.00	23.00		

   a. 3.05
   b. 4.12
   c. 2.18
   d. 1.36

5. Using the following ANOVA source table, compute the Tukey HSD at $\alpha = .01$.

	SS	df	MS	F
Between	89.00	2.00	44.50	9.77
Within	123.00	27.00	4.56	
Total	212.00	29.00		

   a. 2.63
   b. 2.12
   c. 4.12
   d. 5.08

6. Using the following table of means and an HSD of 3.1, which of the following is true?

	Group 1	Group 2	Group 3	Group 4
Mean	8.6	6.8	9.9	11.2

a. The only significant difference is between Groups 2 and 3
b. Group 1 and Group 4 are significantly different
c. Group 2 is significantly different from Group 1 but not from Group 4
d. Group 2 is significantly different from Groups 3 and 4, but not from Group 1

7. A research study is comparing 5 different treatments to a control group. This study has enrolled 120 participants who were randomly assigned to one of the conditions so that there were an equal number of participants in each group. What is q when computing the Tukey HSD for this post hoc test at $\alpha = .05$?
a. 2.35
b. 3.98
c. 5.57
d. 3.12

8. Using the following table, compute a Tukey HSD at $\alpha = .05$ (there are an equal number of participants in each group).

	SS	df	MS	F
Between	367.00	4.00	91.75	5.09
Within	541.00	30.00	18.03	
Total	908.00	34.00		

a. 8.98
b. 12.31
c. 6.90
d. 4.36

9. Using the Tukey HSD from the previous question, use the means provided below to determine which of the following statements are true.

	Group 1	Group 2	Group 3	Group 4	Group 5
Mean	3.45	9.87	4.6	10.8	11.7

a. Groups 1 and 3 are significantly different
b. Groups 2 and 4 are not significantly different, but both are significantly different from Group 1
c. The only significant differences are between Groups 1 and 4 and Groups 1 and 5
d. None of these groups are significantly different

10. An ANOVA compared 3 groups and contained a total of 15 participants. When conducting a Scheffé test, which degrees of freedom should you use when looking up the F critical?
a. 2, 14
b. 2, 12
c. 3, 15
d. 3, 12

Part XI Study Guide

11. The following are the results of a Scheffé test that contained an $MS_{within} = 112$, 3 groups, and a total of 60 participants. Which groups appear significantly different when $\alpha = .05$?

	SS	MS	F
1 and 2	1297.92	432.64	3.86
1 and 3	499.23	166.41	1.49
2 and 3	2170.83	723.61	6.46

    a. 1 and 2
    b. 1 and 3
    c. 2 and 3
    d. a and c

12. The following are the results of a research trial for a new cholesterol medication. The groups were new medication, placebo group, and control group. If there were a total of 45 participants with an equal number of participants in each group, then what values would you use to find q when $\alpha = .05$?
    a. 3, 45
    b. 3, 41
    c. 3, 42
    d. 2, 42

13. Using the information from the previous question, if $MS_{within} = 74.3$, what is the HSD?
    a. 5.32
    b. 12.32
    c. 8.44
    d. 4.23

14. Using the following mean table and the information from Questions 12 and 13, determine which groups are significantly different.

	Group 1	Group 2	Group 3
Mean	124.5	137.5	140.4

    a. Groups 1 and 3 are significantly different
    b. Groups 1 and 2 are significantly different
    c. Groups 2 and 3 are significantly different
    d. a and b

15. The following table provides the SS of a Scheffé test comparing the number of completed passes made by 4 football teams. If the MS_{within} for the primary ANOVA was 22,578 and there were 60 total players across the teams, what is the F ratio for the comparison between Teams 2 and 3?

	SS	MS
1 and 2	46728.38	15576.13
1 and 3	152960.67	50986.89
2 and 3	193788.48	64596.16
1 and 4	308493.38	102831.13
2 and 4	365461.44	121820.48
3 and 4	249573.62	83191.20

a. 2.86
b. 3.78
c. 6.23
d. 4.53

16. Based on the table in the previous question, which statement is true?
    a. Groups 1 and 2 are significantly different
    b. Group 4 is significantly different from Groups 1 and 2
    c. Group 3 is significantly different from Group 1
    d. Group 4 is significantly different from Groups 3 and 2, but not from Group 1

A study was done to determine the influence of domestic violence on depression. Levels of depression were compared in three groups: physical domestic violence in the home; verbal domestic violence in the home; and a nonviolence control group. There were 27 total participants with an equal number of participants in each group. Use the following to answer Questions 17 to 20.

	SS	df
Between	748.00	2.00
Within	897.00	24.00
Total	1645.00	26.00

	Physical	Verbal	Control
Mean	14.00	12.00	4.00

17. In doing a Scheffé test, what is the value of MS_{within}?
    a. 37.38
    b. 45.78
    c. 23.12
    d. 6.37

18. What would be the Tukey HSD for this test at $\alpha = .05$?
    a. 6.32
    b. 7.94
    c. 8.97
    d. 3.23

19. Using the following Scheffé table (from the same data as the current example), which groups appear to be significantly different?

	SS	MS
1 and 2	222.44	74.15
1 and 3	1165.21	388.40
2 and 3	20.07	6.69

a. Groups 1 and 2
b. Groups 2 and 3
c. Groups 1 and 3
d. All groups are significantly different

20. Which groups are significantly different, based on a Tukey HSD test?
    a. Groups 1 and 2 are different from Group 3
    b. Groups 2 and 3 are different from Group 1
    c. Groups 1 and 3 are different from Group 2
    d. All groups are significantly different

Answers to Odd-Numbered Multiple-Choice Questions

1. a
3. d
5. a
7. b
9. c
11. d
13. c
15. a
17. a
19. c

Part XII Study Guide

More Than One Independent Variable

Part XII Summary

Module 28

- ANOVA has two general uses: (1) to examine group differences among three or more groups and (2) to assess differences among multiple independent variables (IVs; regardless of how many groups are associated with each IV). In the second case, the ANOVA is called a *factorial ANOVA* because it has multiple factors (IVs).
- ANOVAs with two IVs are referred to as *two-way ANOVAs*. ANOVAs can have many more factors, with the analysis becoming increasingly more complex with the addition of more factors. Two-way ANOVAs are written as the number of groups in the first IV × the number of groups in the second IV. For example, a study with 3 groups on the first IV and 4 groups on the second IV would be expressed as a 3 × 4 ANOVA.
- Factorial ANOVA enables you to assess numerous hypotheses at once. The first type of hypothesis addresses whether an IV had a significant effect on the DV. These types of hypotheses are referred to as main effects. There is one main effect per IV in the study. Main effects state that there is an effect of one IV on the DV, regardless of any other IVs. For example, assume you were conducting a study on makeup sales and your factors were brand of makeup (3 types) and specific product (3 types). A main effect for brand would indicate that regardless of what type of product, one particular brand is selling more than others. A main effect for product would indicate that regardless of brand, one particular product is selling more than the others. When graphing the group means of main effects, you can expect the lines created by connecting the means to appear parallel.
- The second type of hypothesis addressed by a factorial ANOVA is referred to as an interaction. Interactions indicate that a score on the DV depends on both the first and the second IV. Returning to our makeup example, a significant interaction would indicate that a particular brand/product combination was selling more than any other brand/product combination. When graphing the group means of an interaction effect, you can expect the lines created by the connecting means to intersect.
- It should be noted that using graphs or tables to determine significant main effects and interactions gives an estimate of the probable effect, but it is inexact. The more exact method for assessing significant differences is to calculate the F statistics for main and interaction effects. This will be discussed in the next module.

Module 29

- When calculating a factorial ANOVA from raw data, it is important to be organized because of the tremendous number of calculations.

- The first step when dealing with a factorial ANOVA is to calculate the cell means. Cell means refer to the average score of those in a particular group combination (such as Group 1 of Factor 1 and Group 1 of Factor 2). Next, the row and column means should be calculated. These are the means for all combined groups in a factor. This indicates that the row mean for Factor 1 would be the mean of all the scores in Factor 1, regardless of the group on Factor 2. The cell means are related to *interaction effects*, whereas the row and column means are related to the main effects.
- The *SS* for a factorial ANOVA is calculated in a very similar manner to that of the one-way ANOVA. The purpose of finding the *SS* is to partition it into different sections: variability related to differences between the groups and variability related to random error (within the groups). The variability between the groups can be further broken down into variability related to differences on Factor 1, variability related to differences on Factor 2, and variability related to the interaction. The formulas for the total *SS* are

$$\text{Deviation method: } SS_{tot} = \sum_{1}^{N} (X - M_{tot})^2$$

$$\text{Raw score method: } SS_{tot} = \sum_{1}^{N} X^2 - \frac{\left(\sum X_{tot}\right)^2}{N}$$

The formulas for the within *SS* are

$$\text{Deviation method: } SS_{with} = \sum_{1}^{k} \sum_{1}^{n} (X - M_{cell})^2$$

$$\text{Raw score method: } SS_{with} = \sum_{1}^{N} X^2 - \sum_{1}^{k} \frac{\left(\sum X_{cell}\right)^2}{n_{cell}}$$

The formulas for the between *SS* are as follows (note, formulas for additional factors are the same):

$$\text{Factor 1, deviation method: } SS_{Factor\,1} = \sum_{1}^{k} (M_{Factor\,1} - M_{tot})^2$$

$$\text{Factor 1, raw score method: } SS_{Factor\,1} = \sum_{1}^{k} \frac{\left(\sum X_{Factor\,1}\right)^2}{n_{Factor\,1}} - \frac{\left(\sum X_{tot}\right)^2}{N}$$

The formulas for the interaction term are as follows:

$$SS_{interaction} = SS_{tot} - (SS_{with} + SS_{Factor\,1} + SS_{Factor\,2})$$

- After the *SS* has been partitioned, the next step is to calculate the appropriate degrees of freedom. Remember that the degrees of freedom are used in calculating the mean squares.

$df_{tot} = N - 1$
df_{with} = (No. of groups in Factor 1)(No. of groups in Factor 2)$(n_{cell} - 1)$
$df_{Factor1}$ = No. of groups in Factor 1 − 1
$df_{Factor2}$ = No. of groups in Factor 2 − 1
$df_{interaction}$ = (No. of groups in Factor 1 − 1)(No. of groups in Factor 2 − 1)

- The next step is to calculate the mean squares, which have similar formulas to those of the one-way ANOVA:

$$MS_{with} = \frac{SS_{with}}{df_{with}}; \quad MS_{Factor1} = \frac{SS_{Factor1}}{df_{Factor1}};$$
$$MS_{Factor2} = \frac{SS_{Factor2}}{df_{Factor2}}; \quad MS_{interaction} = \frac{SS_{interaction}}{df_{interaction}}$$

- The final step is to calculate the F ratios. In a factorial ANOVA, there are multiple F ratios, one for each main effect and one for each interaction term.

$$F_{Factor1} = \frac{MS_{Factor1}}{MS_{with}}; \quad F_{Factor2} = \frac{MS_{Factor2}}{MS_{with}}; \quad F_{interaction} = \frac{MS_{interaction}}{MS_{with}}$$

- After obtaining the F ratios, the next step is to determine if they are significant (are the group differences a lot or a little?). This is done in a similar manner to that of a one-way ANOVA by using the table found in Appendix D. Since you have obtained three F ratios, you will need to find three different critical values.
- Similar to a one-way ANOVA, the results can be easily displayed in a source table. Notice that in a factorial ANOVA, there are additional rows that are added to the table.

Source	SS	df	MS	F
Factor 1	$SS_{Factor1}$	$df_{Factor1}$	$SS_{Factor1}/df_{Factor1}$	$MS_{Factor1}/MS_{within}$
Factor 2	$SS_{Factor2}$	$df_{Factor2}$	$SS_{Factor2}/df_{Factor2}$	$MS_{Factor2}/MS_{within}$
Interaction	$SS_{interaction}$	$df_{interaction}$	$SS_{interaction}/df_{interaction}$	$MS_{interaction}/MS_{with}$
Within	SS_{within}	df_{within}	SS_{within}/df_{within}	
Total	SS_{tot}	df_{tot}		

Learning Objectives

Module 28

- Express a factorial ANOVA design in symbols
- Determine the number of possible main effects in an ANOVA
- Determine the number of possible interaction effects in an ANOVA
- From a graph of cell means, determine whether or not there are probable main effects or interaction effects
- From a table of cell means, determine whether or not there are probable main effects or interaction effects

Module 29

- Understand the similar logic underlying various test statistics
- Determine degrees of freedom
- Calculate a factorial ANOVA
- Use a table to interpret F
- Present results in an ANOVA table
- Report results in APA format

Computational Exercises

1. Using the following graph, what are the main effects and interactions (if any)?

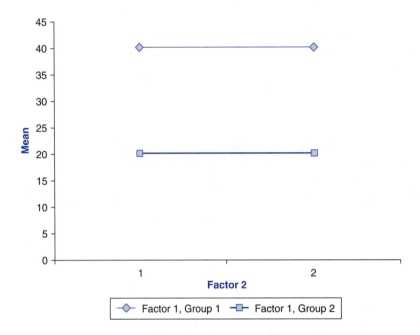

2. Using the following graph, what are the main effects and interactions (if any)?

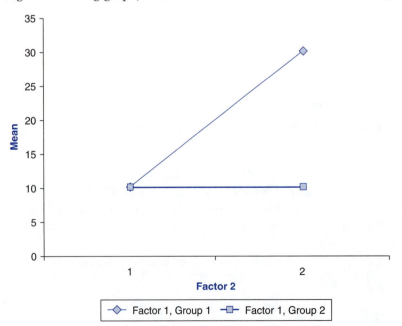

3. Using the following table, what are the main effects and interactions (if any)?

		Factor 2		
		Group 1	Group 2	Group 3
Factor 1	Group 1	10	30	20
	Group 2	10	30	20
	Group 3	10	30	40

4. Using the following table, what are the main effects and interactions (if any)?

		Factor 2	
		Group 1	Group 2
Factor 1	Group 1	45	45
	Group 2	45	45

5. Partition the SS of the following groups:

	Factor 1	
Factor 2 Group	Group 1	Group 2
1	61	62
1	15	49
1	72	12
1	53	9
2	49	31
2	81	85
2	1	73
2	20	74

6. Partition the SS of the following groups:

	Factor 1		
Factor 2 Group	Group 1	Group 2	Group 3
1	76	46	69
1	64	79	45
1	44	42	77
1	75	42	45
2	68	43	56
2	60	69	53
2	50	60	48
2	40	44	72
3	66	56	73
3	66	74	61
3	55	63	40
3	69	76	66

7. A study examines a new treatment for social anxiety by comparing the new treatment to an established treatment and to a separate control group (three groups in total). After the study has been conducted, the therapists anecdotally report that they noticed females seemed to respond better than males. Using the data from the study's primary outcome measure (provided below), design an analysis to address this situation.

	Treatment		
Gender	New Treatment	Established Treatment	Control
Male	2	9	14
Male	6	8	19
Male	3	5	16
Male	8	3	22
Male	7	3	24
Female	5	9	19
Female	7	7	25
Female	5	5	10
Female	5	12	15
Female	9	7	17

a. What are the null and alternative hypotheses?
b. What are the main effects and interactions (if any) at the .05 error level?
c. Present your findings in APA format.

8. Within the diagnosis of schizophrenia, there are two subtypes (positive symptoms, negative symptoms). A new research study seeks to compare the effectiveness of four separate treatments for schizophrenia (SSRI drugs; antipsychotic drugs; therapy; control) across these two subtypes. Using the data below of the rated judgment of change in the client's symptoms, design an analysis that would address the primary question of this study: Does treatment effectiveness vary as a function of schizophrenia subtype?

	Treatment			
Subtype	SSRI	Antipsychotic	Therapy	Control
Positive	1	5	13	13
Positive	10	7	12	19
Positive	1	5	10	19
Positive	1	6	18	17
Positive	10	10	20	9
Positive	6	6	13	11
Negative	5	8	19	20
Negative	10	6	7	15
Negative	4	8	20	7
Negative	6	4	12	7
Negative	9	2	9	10
Negative	8	5	15	11

a. What are the null and alternative hypotheses?
b. What are the main effects and interactions (if any) at the .05 error level?
c. Present your findings in APA format.

9. You are hired by an ice cream maker to develop a new ice cream flavor. Ice cream consists of a base flavor (the actual ice cream) and candy that is mixed in. You decide to assess chocolate and vanilla as your base flavors. You include chocolate chips and peanut butter cups as your mix-ins. You give a group of $N = 108$ people one of each of the four combinations and rank their opinion of the flavor they tasted on a scale of 1 to 10. The company wants to know what combinations you could recommend based on your research.

	Vanilla	Chocolate	Flavor Mean
Choc chips	5.1	6.2	5.65
PB cups	9.76	6.21	7.99
Mix-in M	7.43	6.21	

Source	SS	df	MS	F
	450			
	125			
	80			
	785			

a. Complete the above source table.
b. What are the main effects and interactions (if any) at the .01 error level?
c. Present your findings in APA format.

10. A jeweler is curious as to what women prefer in regard to their engagement rings. She decides to obtain ratings (1–10) from 32 different women (an equal number in each group) on their opinion of their wedding ring in terms of the shape of the stone and the type of metal in the ring setting. Here are the raw scores of satisfaction she obtained from women who own each type of ring.

	Setting	
Shape	Gold Setting	Platinum Setting
Princess	6	8
Princess	10	3
Princess	10	9
Princess	10	5
Princess	8	4
Princess	10	7
Princess	7	7
Princess	9	3
Pear	10	5
Pear	3	6
Pear	9	7
Pear	9	6
Pear	10	5
Pear	9	10
Pear	6	5
Pear	7	3

a. What are the null and alternative hypotheses?
b. What are the main effects and interactions (if any) at the .05 error level?
c. Present your findings in APA format.

11. Two music artists have recently left their respective groups to pursue solo careers. Their record label is concerned about how their fans will perceive this switch and asks four groups of 10 people each their opinion on either one of the groups or the prospect of seeing either artist in a solo career (no one is asked more than one question for fear that rumors might spread). Preference is rated on a 1 to 20 scale. Here are the data the company obtained:

	Solo	Group	Means
Artist A	10.8	6.3	8.55
Artist B	5.4	6.1	5.80
Means	8.1	6.2	

Source	SS	df	MS	F
Between treatments	327			
Solo or group	89			
Artist	122			
Solo × artist				
Within treatments				
Total	789			

a. Complete the above source table.
b. What are the main effects and interactions (if any) at the .01 error level?
c. Present your findings in APA format.

12. The jeweler from Question 10 has just received a large shipment of emeralds, rubies, and sapphires. She has the opportunity to turn any of these stones into earrings, rings, or bracelets. However, she wants to ensure that she does the largest amount of business possible with the pieces she creates. As a result, she reviews the number of sales of each combination of stone and piece from $N = 25$ (an equal number of stores per group) different nearby jewelers. Here are the data she obtains:

	Stone		
Shape	Ruby	Emerald	Sapphire
Earring	11	10	6
Earring	9	21	21
Earring	23	10	15
Earring	18	14	21
Earring	20	7	10
Ring	18	19	21
Ring	5	14	24

		Stone	
Shape	Ruby	Emerald	Sapphire
Ring	6	10	7
Ring	25	15	24
Ring	21	3	14
Bracelet	22	13	15
Bracelet	13	22	7
Bracelet	24	21	15
Bracelet	11	11	8
Bracelet	8	22	23

  a. What are the null and alternative hypotheses?
  b. What are the main effects and interactions (if any) at the .05 error level?
  c. Present your findings in APA format.

13. You are interested in assessing how different environments affect college drinking. You also hypothesize that having an alcoholic parent will interact with the environment to affect the number of drinks consumed. You conduct a study observing the number of drinks consumed in three different college settings: a dorm room, a house party, and a bar. You are able to successfully observe $N = 30$ college students; $n = 10$ different students are observed in each setting. Half of the total sample had an alcoholic parent. You obtain the following descriptive statistics:

	Dorm	House Party	Bar	
Alcoholic parent	2.4	10.2	12.8	8.47
Nonalcoholic parent	2.1	12.3	15.4	9.93
	2.25	11.25	14.1	

Source	SS	df	MS	F
	587	2		
	70			
	487			
Total	1,257			

  a. Complete the above source table.
  b. What are the main effects and interactions (if any) at the .05 error level?
  c. Present your findings in APA format.

14. Three new teachers are interested in determining how well students respond to their different teaching techniques: rewarding, punitive, and experiential. They decide to compare the final grades of the $n = 6$ students in each of their classes to determine which teacher has the most effective technique. One of the teachers suggests that the gender of the student may have an impact on his or her 152 response to the technique. They agree to incorporate this into the study as well. Here are the final grades that they obtained. Is one teaching technique more effective and does gender play a role? Use a .05 error level and present your findings in APA format.

	Technique		
Gender	Rewarding	Punitive	Experiential
Male	90	86	87
Male	89	89	87
Male	81	84	92
Male	81	90	92
Male	79	75	94
Male	84	90	92
Female	77	78	90
Female	84	87	92
Female	84	85	90
Female	82	88	95
Female	75	79	98
Female	90	86	98

15. In conducting research on treatment, it is common to use multiple therapists. However, this could be a potential problem as different therapists may have varying levels of effectiveness. A research study on depression is comparing interpersonal therapy and cognitive behavioral therapy. The researchers are using two therapists and want to ensure that participants are responding to both therapists in a similar manner. Using the following data, determine if there is any difference between the therapists and how (if at all) it is affecting the type of therapy used to treat depression. The scores represent a self-report measure of depression.

	Treatment	
Therapist	Interpersonal	Cognitive-Behavioral
Therapist A	17	8
Therapist A	6	5
Therapist A	10	20
Therapist A	23	12
Therapist A	17	12
Therapist A	6	13
Therapist B	20	23
Therapist B	17	17
Therapist B	25	11
Therapist B	16	10
Therapist B	5	13
Therapist B	17	15

Answers to Odd-Numbered Computational Exercises

1. There is a main effect for Factor 1, but not for Factor 2. There is no interaction.

3. There are two significant main effects and a significant interaction. The third group in Factors 1 and 2 is significantly higher than the rest.

5. $SS_{total} = 1187.44$; $SS_{Factor\ 1} = 410.6$; $SS_{Factor\ 2} = 115.63$; $SS_{interaction} = 2047.56$; $SS_{within} = 9314.25$

7.
   a. Null hypothesis: Treatment and gender are not related to changes in anxiety scores. Alternative hypothesis: Treatment and gender are related to changes in anxiety scores.

	SS	df	MS	F
Treatment	942.20	2.00	471.10	39.31
Gender	2.13	1.00	2.13	0.18
Interaction	22.87	2.00	11.44	0.95
Within	287.60	24.00	11.98	
Total	1254.80	29.00		

   b. There is a significant main effect for treatment. However, there is no main effect for gender, and there is no significant interaction.
   c. There is a significant main effect for treatment, $F(2, 24) = 39.31$, $p < .05$. There is no significant main effect for gender, $F(1, 24) = 0.18$, $p > .05$. There is no significant interaction, $F(2, 24) = 0.95$, $p > .05$.

9.
   a.

Source	SS	df	MS	F
Between treatments	450	3		
Flavor	125	1	125.0	38.81
Mix-in	80	1	80.0	24.84
Flavor × mix-in	245	1	245.0	76.06
Within treatments	335	104	3.2	
Total	785	107		

   b. There are two significant main effects and a significant interaction.
   c. There was a significant main effect for base flavor, $F(1, 104) = 38.81$. There was also a significant main effect for mix-in, $F(1, 104) = 24.84$. These main effects were qualified by a significant interaction, $F(1, 104) = 76.06$.
   The interaction suggested that the most preferred combination was vanilla ice cream with peanut butter cups mixed in.

11.
   a.

Source	SS	df	MS	F
Between treatments	327	3		
Solo or group	89	1	89.0	6.94
Artist	122	1	122.0	9.51
Solo × artist	116	1	116.0	9.04
Within treatments	462	36	12.8	
Total	789	39		

   b. There are two main effects and one significant interaction.
   c. There was a significant main effect for individual solo or group, $F(1, 36) = 6.94$, $p < .01$. There was also a significant main effect for artist, $F(1, 36) = 9.51$, $p < .01$. These main effects were qualified by a significant interaction, $F(1, 36) = 9.04$, $p < .01$.

13.

a.

Source	SS	df	MS	F
Between treatments	770	5		
Environment	587	2	293.5	14.46
Alcoholic parent	70	1	70	3.45
Environment × AP	113	2	56.5	2.784
Within treatments	487	24	20.29	
Total	1,257	29		

b. There is one significant main effect for environment.

c. There was a significant main effect for environment, $F(2, 24) = 14.46$, $p < .05$. There was no significant main effect for having an alcoholic parent, $F(1, 24) = 3.45$, $p > .05$. There was no significant interaction, $F(2, 24) = 2.78$, $p > .05$.

15.

	SS	df	MS	F	Sig
Therapy	16.67	1.00	16.67	0.48	< .05
Therapist	66.67	1.00	66.67	1.91	< .05
Interaction	0.17	1.00	0.17	0.00	< .05
Within	698.33	20.00	34.92		
Total	781.83	23.00			

It appears there was no significant main effect for therapy type, $F(1, 20) = 0.48$, $p > .05$. There was also no significant main effect for therapist, $F(1, 20) = 1.91$, $p > .05$. There was also no significant interaction, $F(1, 20) = 0.17$, $p > .05$. It appears that both therapies were equally effective and both therapists performed in a similar manner.

True/False Questions

1. A study that is examining the influence of hearing damage is investigating the effect of volume of sound and the proximity of the sound to the ear. This would be a factorial ANOVA.

2. In a 2 × 2 factorial ANOVA with no significant interaction but two significant main effects, you would need to do two post hoc tests to determine the significant differences among the main effects.

3. In a 3 × 3 × 2 ANOVA, the within-subject variability is divided into four sections.

4. An interaction indicates that the effect of one IV on the DV is conditional on the level of a second IV.

5. A main effect is similar to the results found in a one-way ANOVA.

6. When the row and column means are identical, there is a significant interaction.

7. When graphing the means, overlapping lines indicate a lack of all main effects.

8. When graphing the means, separate but parallel lines indicate a main effect for that factor and a lack of an interaction.

9. In real research, you can expect perfectly parallel lines in the absence of an interaction.

10. An ANOVA with three groups on one factor, two groups on the second factor, and four groups on the third factor could be written as a 24-way ANOVA.

11. There are no interactions in the following ANOVA:

		Factor 2		
		Group 1	Group 2	Group 3
Factor 1	Group 1	10	15	15
	Group 2	10	15	15
	Group 3	10	15	15

12. In a factorial ANOVA, all the between-treatment variabilities are compared with the same random error component.

13. The degrees of freedom used to determine the critical value can change depending on which main effect and interaction you are assessing.

14. A 2 × 2 factorial ANOVA has 32 participants in total (with an equal number of participants in each group). The correct degrees of freedom for the interaction are 1 and 29.

15. A significant interaction may indicate that the highest group mean is different from that indicated by your main effects.

Answers to Odd-Numbered True/False Questions

1. True
3. False
5. True
7. False
9. False
11. True
13. True
15. True

Short-Answer Questions

1. When is it appropriate to use a factorial ANOVA?
2. What is a main effect?
3. How does an interaction differ from a main effect?
4. Create a graph for a significant 2 × 3 interaction effect.
5. Is it possible to create a graph for a three-way interaction? Why or why not?
6. Why is relying solely on a table or graph not the best method for assessing significant differences in a factorial ANOVA?
7. How many interaction effects are possible in a four-way ANOVA?
8. Where should you inspect on a table of means for a significant interaction effect and a significant main effect?

9. In a three-way ANOVA, how many F ratios should you expect to calculate? What are they?
10. In a 3 × 6 ANOVA, how many different groups can you expect to compare in the main effects and the interaction?
11. An ANOVA includes three F tests, each with its own critical value. Does this mean that an ANOVA is increasing the chance of committing a Type 1 error?
12. In a study of career enjoyment, a researcher is interested in determining how people feel about their careers. The researcher hypothesizes that the length in which one is involved with a certain career influences one's satisfaction. Due to the gender gap in salary and promotion, the researcher wants to incorporate gender into the analysis. How many IVs are in this study?
13. Suggest a study for a 3 × 3 ANOVA.
14. Suggest a study for a 3 × 3 × 2 ANOVA.

Answers to Odd-Numbered Short-Answer Questions

1. Factorial ANOVAs are used when there are multiple IVs.

3. An interaction differs from a main effect in that an interaction looks for significant differences using both the factors. Specifically, the influence of one IV on the DV is dependent on the influence of a second IV.

5. It is not possible to graph such an interaction in two dimensions. This is because you have no way of showing the effect of the third IV. Alternative methods include creating a 3-D graph or creating multiple graphs. The second method entails creating separate graphs (showing the groups of the third IV) and then within each graph, showing the interactions and main effects of the first two IVs.

7. There will be four separate main effects (one for each IV) and 11 interaction effects.

9. There will be seven F ratios: three main effects and four interactions.

11. No, this is not the case. This is because of the shape of the F distribution and the number of degrees of freedom that are used in determining the critical value.

13. Answers will vary.

Multiple-Choice Questions

1. In the following table, there are

		Factor 1	
		Group 1	Group 2
Factor 2	Group 1	50	40
	Group 2	90	80

   a. Two main effects
   b. A main effect for Factor 1 but not for Factor 2
   c. Two main effects and an interaction
   d. An interaction

2. In the following table, you would find

		Factor 1	
		Group 1	Group 2
Factor 2	Group 1	10	15
	Group 2	10	15

   a. Two main effects
   b. A main effect for Factor 1 but not for Factor 2
   c. Two main effects and an interaction
   d. An interaction

3. What would X need to be in order for there to be a main effect for Factor 2 and no interaction?

		Factor 1	
		Group 1	Group 2
Factor 2	Group 1	10	45
	Group 2	20	X

   a. 45
   b. 20
   c. 30
   d. 55

4. How many pieces will the between-subject variability be divided into in a 3 × 4 ANOVA?
   a. 3
   b. 4
   c. 12
   d. 7

5. A 2 × 2 study contains $N = 40$ people evenly distributed across the groups. What are the degrees of freedom for this study?
   a. $df_{total} = 39; df_{Factor\ 1} = 2; df_{Factor\ 2} = 2; df_{interaction} = 4; df_{within} = 31$
   b. $df_{total} = 39; df_{Factor\ 1} = 1; df_{Factor\ 2} = 1; df_{interaction} = 2; df_{within} = 35$
   c. $df_{total} = 39; df_{Factor\ 1} = 1; df_{Factor\ 2} = 1; df_{interaction} = 1; df_{within} = 36$
   d. $df_{total} = 40; df_{Factor\ 1} = 1; df_{Factor\ 2} = 1; df_{interaction} = 1; df_{within} = 36$

6. In the following table, which are significant?

	SS	df	MS	F
Factor 1	25.00	1.00	25.00	1.76
Factor 2	12.00	1.00	12.00	0.85
Interaction	14.00	1.00	14.00	0.99
Within	213.00	15.00	14.20	
Total	264.00	18.00		

   a. The main effects but not the interaction
   b. The main effects and the interaction
   c. The interaction but not the main effects
   d. None are significant

7. For the main effect for Factor 1 in the previous table, which of the following is the correct presentation of the result in APA format?
   a. $F(1, 15) = 1.76, p < .05$
   b. $F(1, 18) = 1.76, p < .05$
   c. $F(1, 15) = 1.76, p > .05$
   d. $F(1, 18) = 1.76, p > .05$

8. What would X need to be in order for there to be no main effects?

		Factor 1	
		Group 1	Group 2
Factor 2	Group 1	5	30
	Group 2	30	X

   a. 30
   b. 5
   c. 10
   d. 25

9. In the following graph, how many interactions and main effects are there?

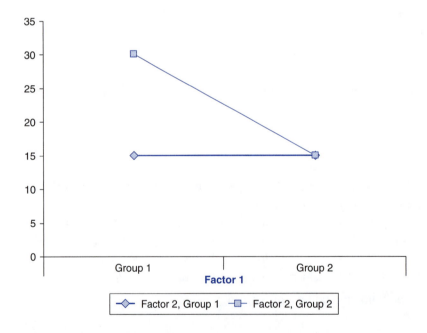

   a. The main effects, but not the interaction are significant
   b. The main effects and the interaction are significant
   c. The interaction but not the main effects are significant
   d. None are significant

10. When graphing means, what do two parallel (or close to parallel) lines indicate?
    a. The main effects but not the interaction
    b. The main effects and the interaction
    c. The interaction but not the main effects
    d. At least one main effect is significant

11. If all the row and column means of a mean table are the same, then what is significant?
    a. The main effects but not the interaction
    b. The main effects and the interaction
    c. The interaction but not the main effects
    d. At least one main effect is significant

12. A study was conducted on the influence of supervision on improving the skills of a therapist. The study was interested in determining how theoretical orientation of the supervisor and the length of time of the supervision (1 week, 2 weeks, 3 weeks, 4 weeks) influenced therapist improvement. Here is a table of the results:

		\multicolumn{4}{c}{Time in Weeks}
		1	2	3	4
Orientation	CBT	20	34	47	45
	Interpersonal	21	29	41	44

    Based on this table, what appears to be significant?

    a. The main effects but not the interaction
    b. The main effects and the interaction
    c. The interaction but not the main effects
    d. At least one main effect is significant

13. Here is the source table for the example in Question 12. Which effects are significant?

	SS	df
Time	250.00	1.00
Orientation	210.00	4.00
Interaction	124.00	4.00
Within	689.00	40.00
Total	1,273.00	49.00

    a. The main effect for time but not orientation
    b. The main effect for orientation but not time
    c. Both main effects
    d. The interaction

14. Based on the previous question, what are the results of this study?
    a. It appears that increasing supervision time improves the skills of a therapist
    b. It appears that decreasing supervision time improves the skills of a therapist
    c. It appears that orientation does not play a role in increasing therapy skills
    d. It appears that those in CBT supervision improve faster than those in interpersonal supervision

15. Using the table from Question 13, how many participants were involved in the study?
    a. 49
    b. 50
    c. 51
    d. 55

16. Using the following table, which F ratio(s) are significant?

	SS	df	MS
Factor 1	32.00	1.00	32.00
Factor 2	123.00	4.00	30.75
Interaction	245.00	4.00	61.25
Within	310.00	35.00	8.86
Total	710.00	44.00	

a. Factor 1
b. Factor 2
c. The interaction
d. All the F ratios are significant

17. Using the following table, what are the mean squares?

	SS	df
Factor 1	450.00	2.00
Factor 2	345.00	3.00
Interaction	987.00	2.00
Within	1,247.00	18.00
Total	3,029.00	25.00

a. $MS_{Factor\ 1} = 225$; $MS_{Factor\ 2} = 115$; $MS_{interaction} = 493.5$; $MS_{within} = 69.28$
b. $MS_{Factor\ 1} = 50$; $MS_{Factor\ 2} = 65$; $MS_{interaction} = 214.5$; $MS_{within} = 632$
c. $MS_{Factor\ 1} = 127$; $MS_{Factor\ 2} = 3,214$; $MS_{interaction} = 1,246$; $MS_{within} = 5,434$
d. $MS_{Factor\ 1} = 324$; $MS_{Factor\ 2} = 654$; $MS_{interaction} = 324$; $MS_{within} = 987$

18. Using the table in Question 17, which F ratio(s) are significant?
a. Factor 1
b. Factor 2
c. The interaction
d. All the F ratios are significant

19. Without altering the SS or the Type 1 error level, how could a researcher increase the chances that he or she would obtain a significant result?
a. Include additional IVs
b. Increase the sample size
c. Select another sample
d. There is nothing the researcher could do

20. Which (if any) may be significant in the following graph?

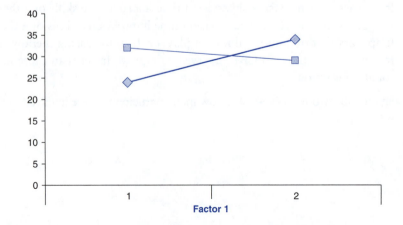

a. Two main effects
b. An interaction
c. One main effect
d. a and b

Answers to Odd-Numbered Multiple-Choice Questions

1. a
3. d
5. c
7. c
9. b
11. c
13. a
15. b
17. a
19. b

Part XIII Study Guide

Nonparametric Statistics

Part XIII Summary

Module 30

- Tests that are grounded in the probabilities for populations are referred to as *parametric tests*. However, in certain cases, such as using small samples or dealing with highly skewed data and data that are not based on an interval scale, it may be more useful to use a nonparametric test. *Nonparametric tests* are not based on population parameters and are thus able to navigate the previously mentioned situations.
- A *chi-square* test assesses the statistical significance of frequencies in different nominal categories. In other words, a chi-square test assesses if one group is similar to another group. A one-variable chi-square test is also referred to as a *goodness-of-fit test*. The test is determining how well the data fit your expected frequencies across the groups.
- The logic of the chi-square test follows the logic of all the previous hypothesis tests. It compares the difference between the obtained group frequencies with the expected frequencies. The formula for a chi-square test is as follows:

$$\chi^2 = \sum \frac{(f_o - f_e)^2}{f_e}$$

- The difficult part of a chi-square test is obtaining the correct expected frequencies. For this section, the expected frequencies will be provided. However, in a real-world situation it can be difficult to ground your expectations of frequencies in empiricism.
- After obtaining your chi-square value, you must use the table found in Appendix F to determine if the difference between the expected and obtained frequencies is significant. The chi-square test can only be one-tailed, as is reflected by the table. This is because a chi-square statistic is always a positive value. The *df* for a chi-square is equal to the number of categories in the IV − 1.
- Significant chi-squares are expressed as the following:

$$\chi^2(df) = \text{chi-square value}, p \text{ value}$$

Module 31

- Similar to the manner in which an ANOVA can assess group differences among two independent variables, a chi-square can include a second IV. When a chi-square test contains two independent variables, it is known as a *test of independence*. The purpose of this test is to determine if the first IV is related to or independent of the second IV.

- A test of independence requires that the expected frequency of each cell be at least 5, regardless of the observed frequency. This is because small expected frequencies have similar difficulties to that of small sample sizes in parametric tests; the chances of making an error are increased. Although there are correction factors that can be applied to a chi-square in which an expected frequency is less than 5, you should try to either conceptualize your groups in a different manner or select a different analysis.
- In doing a test of independence, your variables must be independent. Variables that are not independent of one another will influence your results. Imagine trying to determine if enrollment in a specific section of a psychology course is related to the dorms in which students live. You should expect that this decision will occur at random. However, students in the same dorm may provide each other information about the different professors who would unfairly influence the results.
- The formula for the test of independence is the same as that for the test of a one-variable chi-square test. The only difference is that the number of categories is the total number of cells across both variables.
- In conducting a test of independence, first find the expected frequency. This can be obtained from the information provided to you. Alternatively, you can obtain the expected frequencies using a formula by multiplying the total frequency for a column by the total frequency for the row and then dividing by N. After obtaining the expected frequency, the chi-square statistic is then calculated using the same formula as that for the one-variable chi-square. This obtained result is then compared with a value obtained from the table in Appendix F. The df used for this test is the number of columns $-$ 1 multiplied by the number of rows $-$ 1. If the value obtained from the chi-square exceeds the value obtained from the table, then the two variables are not considered to be independent.

Learning Objectives

Module 30

- Understand the similar logic underlying various test statistic
- Determine degrees of freedom
- Calculate a one-variable chi-square
- Use a table to interpret a chi-square
- Report results in APA format

Module 31

- Understand the similar logic underlying various test statistics
- Know the assumptions underlying a chi-square test of independence
- Find the expected frequencies from row and column totals
- Determine degrees of freedom
- Calculate a two-variable chi-square
- Use a table to interpret a chi-square
- Report results in APA format

Computational Exercises

1. A recent nationwide poll suggested that 28% of the nation supports a specific politician. Your local town appears to consist of a different political ideology than the

nation and you want to test this hypothesis. You are able to poll 400 members of your town and find that 132 support the politician.
   a. What are the null and research hypotheses?
   b. Create a table that shows your comparison.
   c. Calculate χ^2.
   d. Interpret your results at a .05 error level.

2. A college has a 7 to 9 male to female ratio. The incoming class has 432 females and 520 males. Does this incoming class differ from that of the college overall?
   a. What are the null and research hypotheses?
   b. Create a table that shows your comparison.
   c. Calculate χ^2.
   d. Interpret your results at a .05 error level.

3. A department store divides its sales year into four parts—Spring, Summer, Fall, and Winter. They usually expect to sell 300 items in the spring, 300 in the summer, 400 in the fall, and 550 in the winter. This year, they have a surprisingly good year and sell 310 in the spring, 330 in the summer, 453 in the fall, and 574 in the winter. Would you classify this as a good year for the store?
   a. What are the null and research hypotheses?
   b. Create a table that shows your comparison.
   c. Calculate χ^2.
   d. Interpret your results at a .05 error level.

4. Data on income across the nation were recently released: 20% earn less than $20 thousand per year, 13% earn between $20 and $40 thousand per year, 32% earn between $40 and $60 thousand per year, 15% earn between $60 and $80 thousand per year, 10% earn between $80 and $100 thousand per year, 5% earn between $100 and $200 thousand per year, and 5% earn more than $200 thousand per year. A major city claims that it is the richest city in the world having 100,000 residents, who fall in different income brackets as follows: 13% earn less than $20 thousand per year, 15% earn between $20 and $40 thousand per year, 16% earn between $40 and $60 thousand per year, 20% earn between $60 and $80 thousand per year, 20% earn between $80 and $100 thousand per year, 10% earn between $100 and $200 thousand per year, and 6% earn more than $200 thousand per year. Are they correct in their claim?
   a. What are the null and research hypotheses?
   b. Create a table that shows your comparison.
   c. Calculate χ^2.
   d. Interpret your results at a .05 error level.

5. A song writer receives a complaint from a guitarist that he has written a song that is too complex to play. The guitarist states that he can normally play a song 85% correctly on his first attempt. With the new song, he could only play 329 of the 813 notes in the song. Is the guitarist's complaint warranted?
   a. What are the null and research hypotheses?
   b. Create a table that shows your comparison.
   c. Calculate χ^2.
   d. Interpret your results at a .01 error level.

6. As the owner of a clothing store, you must purchase new shirts in multiple colors: blue, red, pink, and yellow. You originally want to purchase an equal number (25%) of each color. After ordering 1,000 shirts, you notice that the different colors have sold at

different rates. The blue has sold 120, the red has sold 250, the pink has sold 250, and the yellow has sold 87. Were you correct in ordering an equal number of each color shirt?
   a. What are the null and research hypotheses?
   b. Create a table that shows your comparison.
   c. Calculate χ^2.
   d. Interpret your results at a .01 error level.

7. The head chef is preparing a number of dishes for a banquet of 1,000. He arrived with enough ingredients to prepare for the following food preferences: 10% vegetarians, 35% chicken, 25% beef, 15% pork, and 15% fish. He has purchased additional ingredients in case the preferences change slightly from expectation, but he will need a lot more if his observation is very different from what he has been told. On arriving, he finds out that the requests are as follows: 97 vegetarian, 142 chicken, 320 beef, 210 pork, and 231 fish. Should the chef send out his assistant to purchase additional ingredients?
   a. What are the null and research hypotheses?
   b. Create a table that shows your comparison.
   c. Calculate χ^2.
   d. Interpret your results at a .01 error level.

8. A farmer is somewhat concerned about his apple crop this year. He covered his orchard with 30% Gala apples, 20% Red Delicious, 30% McIntosh, and 20% Empire. However, he harvested a total of 1,049 apples: 312 Gala, 126 Red Delicious, 432 McIntosh, and 179 Empire. Should he be perplexed by his yield?
   a. What are the null and research hypotheses?
   b. Create a table that shows your comparison.
   c. Calculate χ^2.
   d. Interpret your results at a .01 error level.

9. Prior research has indicated that 40% of children are perpetrators of bullying and 60% are victims. A team of researchers is interested in seeing if there is a gender difference between those who are perpetrators and those who are victims. The researchers originally expected there to be no gender divide such that proportions did not differ between boys and girls. In a study on bullying containing 30 participants (15 boys and 15 girls), the researchers notice that 4 of the boys and 6 of the girls are perpetrators of bullying. Does this sample indicate that status as victim or perpetrator is dependent on gender?
   a. What are the null and research hypotheses?
   b. Create a table that shows your comparison.
   c. Calculate χ^2.
   d. Interpret your results at a .05 error level.

10. A doctor notices that many of her patients are teachers with the flu. The doctor wonders if being a teacher is related to having the flu. In the past week, the doctor has seen a total of 20 patients, 4 of whom have been teachers. Also, of the teachers, all 4 have had the flu while only 3 of the other patients have had the flu. Does it appear that being a teacher is related to having the flu?
    a. What are the null and research hypotheses?
    b. Calculate χ^2.
    c. Interpret your results at a .05 error level.

11. A cookbook company is curious about what is the better food to be paired with peanut butter: jelly or bananas. They decide to ask a group of 40 people to rate which they prefer. The company notices that there appear to be two age demographics among those that they asked: half of the sample is between 10 and 15 and the other half is between 30 and 35. The company expected there to be an equal preference

across all groups. If 15 of the younger group prefer jelly and 12 of the older group prefer bananas, does there appear to be a relation between age and preference?
   a. What are the null and research hypotheses?
   b. Calculate χ^2.
   c. Interpret your results at a .05 error level.

12. A study seeks to determine if there is a gender difference for beer and wine preference. The researcher hypothesizes that there will be no difference between the genders. A study was done in which 50 men and 50 women were asked to state their preferred drink. Forty-five men preferred wine and 35 women preferred beer. Does gender appear to be related to alcoholic beverage preference?
    a. What are the null and research hypotheses?
    b. Calculate χ^2.
    c. Interpret your results at a .05 error level.

13. A movie producer is interested in determining if those of different ages prefer action movies or dramas. According to recent demographics, 75% of people below the age of 25 prefer action movies. However, 65% of people above the age of 25 prefer dramas. The producer releases an action movie and discovers that of his movie's first 500 patrons, 312 are below the age of 25. Should this result surprise the movie producer?
    a. What are the null and research hypotheses?
    b. Calculate χ^2.
    c. Interpret your results at a .05 error level.

14. A research group is interested in determining if education is related to political affiliation. They hypothesize that there will be no difference among education status and political party membership. They develop two categories of education, one containing people with a college degree and another containing those without. They ask 100 individuals (50 with a college degree, 50 without) to list their political affiliation (Democrat or Republican). Twenty-three of those with a college degree are Democrats and 27 of those without a college degree are Republicans. Does it appear that education is related to political affiliation?
    a. What are the null and research hypotheses?
    b. Calculate χ^2.
    c. Interpret your results at a .05 error level.

15. There has been consistent research that shows depression rates in women are approximately 10% and that depression is twice as prevalent in women as in men. However, a researcher is skeptical of this. The researcher obtains a sample of 40 men and 40 women and asks them if they have ever been depressed. Eight of the men and 10 of the women state they were depressed at least once. Does it appear that depression is related to gender?
    a. What are the null and research hypotheses?
    b. Create a table that shows your comparison.
    c. Calculate χ^2.
    d. Interpret your results at a .05 error level.

Answers to Odd-Numbered Computational Exercises

1.
   a. Null hypothesis: Your town does not hold a different opinion than that of the nation. Alternative hypothesis: Your town does hold a different opinion than that of the nation.

b.

Support		Does Not Support	
Observed	Expected	Observed	Expected
132	112	268	288

c. $\chi^2 = \frac{(132-112)^2}{112} + \frac{(268-288)^2}{288} = 4.96$

d. Reject the null hypothesis. Your town has a higher level of support for the politician than expected.

3.
a. Null hypothesis: The sales this year were not different from sales of previous years. Alternative hypothesis: The sales this year were different from sales of previous years.

b.

Spring		Summer		Fall		Winter	
Observed	Expected	Observed	Expected	Observed	Expected	Observed	Expected
310	300	330	300	453	400	574	550

c. $\chi^2 = 11.40$

d. Critical $\chi^2 = 7.81$. Reject the null hypothesis. This year, the store sold more items than last year.

5.
a. Null hypothesis: The song is not more complex than prior songs. Alternative hypothesis: The song is more complex than prior songs.

b.

Notes	
Observed	Expected
329	691

c. $\chi^2 = 189.64$

d. Critical $\chi^2 = 3.84$. Reject the null hypothesis. The song is significantly more difficult than other songs.

7.
a. Null hypothesis: There is no difference between the actual requests and expected requests. Alternative hypothesis: There is a difference between the actual requests and expected requests.

b.

Vegetarian		Chicken		Beef		Pork		Fish	
Observed	Expected	Observed	Expected	Observed	Expected	Observed	Expected	Observed	Expected
97	100	142	350	320	250	210	150	231	150

c. $\chi^2 = 211.04$
d. Critical $\chi^2 = 15.09$. Reject the null hypothesis. The requests at the banquet differ significantly from what the chef expected. They should send out their assistant.

9.
   a. There is no significant relation between perpetrator or victim status and gender.
   b.

	Perpetrator	Victim
Boy	4 (6)	11 (9)
Girl	6 (6)	9 (9)

   c. $\chi^2 = 3.11$
   d. Critical $\chi^2 = 7.81$. Retain the null hypothesis. Gender appears to be unrelated to perpetrator or victim status.

11.
   a. Age is independent of preference for peanut butter addition.
   b. $\chi^2 = 5.8$
   c. Critical $\chi^2 = 7.81$. Retain the null hypothesis. Age is unrelated to your preference for peanut butter addition.

13.
   a. 70% of the patrons will be below the age of 25.
   b. $\chi^2 = 42.34$
   c. Critical $\chi^2 = 3.84$. Reject the null hypothesis. Significantly less than 70% of the patrons were above the age of 25.

15.
   a. Depression is independent of gender.
   b.

	Depressed	Not Depressed
Male	8 (2)	32 (38)
Female	10 (4)	30 (36)

   c. $\chi^2 = 28.95$
   d. Critical $\chi^2 = 7.81$. Reject the null hypothesis. Depression is related to gender.

True/False Questions

1. Nonparametric tests do not rely on means, but they do rely on variances.
2. You should consider using a nonparametric test with highly skewed data.
3. Chi-square tests assess significant differences in frequencies of only one category.
4. The purpose of a test of independence is to determine if two variables are related.
5. A goodness-of-fit test determines whether two groups are independent.
6. Chi-square significance is not determined by probability.
7. The logic for the nonparametric tests is different from the logic for the parametric tests.
8. If an IV has 10 categories, the *df* for the chi-square is 10.

9. There are no two-tailed chi-square tests.

10. The chi-square distribution is very similar to the t distribution.

11. A test of independence has four categories on one IV and three categories on the second IV. The df for this study is 6.

12. In a test for independence, the two IVs must be dependent.

13. The chi-square test is not sensitive to small frequencies.

14. Expected frequencies can be difficult to find in real data.

15. The formulas for tests of independence and goodness of fit are the same.

Answers to Odd-Numbered True/False Questions

1. False
3. False
5. False
7. False
9. True
11. True
13. False
15. True

Short-Answer Questions

1. In which situations are parametric tests not appropriate?
2. What is a nonparametric test? What are some of the questions that can be addressed using a nonparametric test?
3. What is the difference between an ANOVA, a t test, and a chi-square test?
4. Explain the logic behind the formula of the chi-square test.
5. Why can it be difficult to discover the expected frequencies?
6. Why are chi-square tests always one-tailed?
7. What piece of a factorial ANOVA is a test of independence most similar to? Why is this?
8. What additional criteria are required for a test of independence?
9. Why must the expected cell frequency in a test of independence be at least 5? How can this be corrected should an expected frequency be less than 5?
10. What is the difference between a test of independence and a test for goodness of fit?
11. A researcher is interested in determining if there is a gender difference in whether or not people play video games. The researcher also hypothesizes that there may be a

role of age such that adolescents play more than adults. Is this a test of independence or goodness of fit?

12. A principal wants to determine if any of the teachers are inflating grades by giving too many grades of an A. Would the test she conducts be a goodness-of-fit test or a test for independence?

13. A researcher wants to determine the age at which language is first present. Design a goodness-of-fit study to assess this hypothesis.

14. An amusement park is interested in determining the size of the families that are coming to their attractions. They count the frequencies of the families depending on how many children they have (1, 2, 3, 4, 5, or greater than 5). Select another categorical variable that the amusement park should consider in their analysis that may or may not be related.

15. A study recently obtained $\chi^2 = 34$ with 7 categories. At a Type 1 error rate of .01, how should this result appear in APA format?

Answers to Odd-Numbered Short-Answer Questions

1. Parametric tests are not appropriate when using data that are skewed, when the sample size is very small, and finally when the data are not on an interval scale.

3. An ANOVA compares the means and variances of multiple categorical groups. A *t* test compares the means and variances of two categorical groups. A nonparametric test compares the frequencies of categorical groups with expected frequencies.

5. Expected frequencies are based on information that is obtained prior to conducting the study. This can be difficult to find in research situations because such information may not be readily available.

7. The interaction portion of the ANOVA. The interaction states that the influence of the IV on the DV is dependent on a second IV. A test of independence states that the frequency of one group is dependent on its group in another variable.

9. This is because small cell values can result in an inflation of the error term when calculating the chi-square. This is similar to how small sample sizes can affect parametric tests.

11. This is a test of independence. There are two variables, age and gender.

13. Answers will vary.

15. $\chi^2(6) = 34, p < .01$.

Multiple-Choice Questions

1. What scale of measurement is most commonly used for the independent variables in a chi-square test?
   a. Nominal
   b. Interval

c. Ordinal
d. Ratio

2. What is the minimum expected cell frequency for a chi-square test?
   a. There is no minimum
   b. 10
   c. 1
   d. 5

3. The shape of the chi-square distribution for significant difference is
   a. Positively skewed
   b. Negatively skewed
   c. Symmetrical
   d. Bimodal

4. What is the smallest possible chi-square value?
   a. 1
   b. 0.5
   c. 0
   d. −1

5. Which of these is a parameter?
   a. Mean
   b. Standard deviation
   c. Variance
   d. All the above

6. You expected 10 scores in Category A and 30 scores in Category B, but obtained 9 scores in Category A and 33 scores in Category B. What is the chi-square for this situation?
   a. $\chi^2 = 0.40$
   b. $\chi^2 = 1.40$
   c. $\chi^2 = 0.30$
   d. $\chi^2 = 1.30$

7. What is the chi-square for the following table?

A		B	
Observed	Expected	Observed	Expected
112	120	125	145

   a. 4.25
   b. 3.29
   c. 6.21
   d. 7.69

8. What is the chi-square for the following table?

A		B		C	
Observed	Expected	Observed	Expected	Observed	Expected
320	335	321	289	214	231

a. 4.25
b. 5.27
c. 3.36
d. 0.14

9. Which is the correct presentation of the results from the following table?

	A		B		C	
	Observed	Expected	Observed	Expected	Observed	Expected
	10	7	12	15	7	8

a. $\chi^2(2) = 12.3, p > .05$
b. $\chi^2(2) = 2.01, p > .05$
c. $\chi^2(3) = 12.3, p > .05$
d. $\chi^2(3) = 2.01, p > .05$

10. What is the chi-square for the following table?

	A		B	
	Observed	Expected	Observed	Expected
C	5	8	10	14
D	7	9	2	5

a. 4.51
b. 5.41
c. 12
d. 3.1

11. Using the table in Question 10, how many degrees of freedom are involved in finding the critical chi-square?
    a. 1
    b. 2
    c. 3
    d. 4

12. A researcher is interested in determining if different subtypes of schizophrenia are related to the number of hallucinations a person has per week. The researcher observes four subtypes of schizophrenics for 10 weeks. What analysis should the researcher use?
    a. χ^2 goodness of fit
    b. χ^2 test of independence
    c. ANOVA
    d. *t* test

13. A researcher is interested in determining if there is an interaction between drinking coffee (any amount) and job type. What test would be most appropriate to assess this question?
    a. χ^2 goodness of fit
    b. χ^2 test of independence
    c. ANOVA
    d. *t* test

14. Below are the results from the study posed in the previous question. What is the chi-square?

	Doctor		Lawyer		Professor	
	Observed	Expected	Observed	Expected	Observed	Expected
Coffee drinker	8	10	7	10	15	10
Non-coffee drinker	12	10	13	10	5	10

   a. 3.2
   b. 4.3
   c. 7.6
   d. 4.1

15. How would you interpret the results of the previous question?
   a. Being a coffee drinker is related to one's occupation
   b. Overall, doctors drink less coffee than lawyers
   c. Professors drink the least amount of coffee
   d. Being a coffee drinker causes one to be a professor

16. For the table in Question 14, how many degrees of freedom are there?
   a. 1
   b. 2
   c. 3
   d. 4

17. A department store has two entrances, a main entrance and a side entrance. The owner suspects that 75% of his customers enter via the main entrance. At the end of the day, he determines that of his 423 customers, 189 entered through the side entrance. Was the owner correct in his assumption?
   a. Yes, he was correct, $\chi^2(1) = 88.93$, $p < .05$
   b. No, he was incorrect, $\chi^2(1) = 88.93$, $p > .05$
   c. Yes, he was correct, $\chi^2(2) = 88.93$, $p < .05$
   d. No, he was incorrect, $\chi^2(2) = 88.93$, $p > .05$

18. For χ^2, larger expected values are related to
   a. A greater chance of finding a significant chi-square
   b. A reduced chance of finding a significant chi-square
   c. It is unrelated to a significant chi-square
   d. It depends on the difference between the observed and expected value

19. A college admissions officer suspects that having a pet is related to becoming a veterinarian. She expects that if this were not true, an equal number of veterinarians would have pets as not have pets. To test this hypothesis, she polls the past 100 graduates that have become veterinarians and asks them if they had ever owned a pet. If she finds that 78 veterinarians have pets, what is the calculated chi-square for this example?
   a. 3.214
   b. 31.36
   c. 24.32
   d. 45.36

20. A colleague of the person in Question 19 suggests that this question would be better assessed by comparing these results to nonveterinarians. The admissions officer agrees

and collects information from 100 nonveterinarians. With this new test, the admissions officer expects that 78 of the veterinarians will own pets and 22 will not (the information from the previous question). The admissions officer suggests that if being a veterinarian does not influence pet ownership, then the same number of nonveterinarians as veterinarians should own pets. What would be the calculated chi-square for this new test?

Pet Owner		Non–Pet Owner	
Observed	Expected	Observed	Expected
64	78	36	22

a. 11.42
b. 23.1
c. 0.25
d. 1

Answers to Odd-Numbered Multiple-Choice Questions

1. a
3. a
5. d
7. b
9. b
11. a
13. b
15. a
17. a
19. b

Part XIV Study Guide

Effect Size and Power

Part XIV Summary

Module 32

- All the statistics that have been covered thus far have enabled us to make a dichotomous decision about our hypotheses, namely whether the observed difference was large or not. However, it is important to ask another question: "Does this difference matter?" This is separate from a statistical difference; as you have seen, very large samples will make even small differences appear statistically significant. The important question for research that statistics hopes to answer is, "Does this treatment work? Is it effective?" It should be made clear that a hypothesis test does not fully answer this question. Measures that do attempt to answer the question "Is this an effective treatment?" are called measures of *effect size*.
- Measures of effect size are largely subjective. They do not provide an outcome that states "This is effective" or "This is not effective." It is up to the researcher to determine what constitutes an effective treatment. Measures of effect size just provide a good starting point.
- One measure of effect size is called a *Cohen's d*, which rescales the difference between two means to standard deviation units. Cohen's d is used for studies comparing two means, such as a two-sample t test. The formula for Cohen's d is

$$d = \frac{M_1 - M_2}{s}$$

- At first glance, the formula for Cohen's d appears very similar to that of an independent-samples t test. However, the formula for Cohen's d uses a standard deviation in the denominator rather than a standard error. By using a standard deviation, Cohen's d is able to compare samples within that particular study.
- There are no set cutoff points for interpreting a Cohen's d. The author of the statistic suggested that 0.2 standard deviations was indicative of a small effect, 0.3 to 0.8 a medium effect, and anything greater than 0.8 was a large effect.
- A second measure of effect size for studies comparing two means is r, which measures the extent to which two variables are related. In terms of effect size, r represents the association between the IV and the DV. This association ranges from –1 to 1 with 0 indicating no relationship. The formula for effect size r is

$$r = \sqrt{\frac{t^2}{t + df}}$$

- A third measure of effect size is *eta* (η). Eta is used for studies of three or more means, such as those tested with an ANOVA. Similar to *r*, this effect size ranges from 0 to 1. This formula asks you to determine the proportion of between-treatment variability to total variability. In other words, it asks the question, "How much of the total variation is attributed to differences among the groups?" The formula is as follows:

$$\eta = \sqrt{\frac{SS_{bet}}{SS_{tot}}}$$

- For a chi-square test, the measures of effect size vary depending on which test you have conducted, a goodness-of-fit test or a test of independence. For a goodness-of-fit test, effect size is calculated with the following formula:

$$\text{Effect size} = \sqrt{\frac{\chi^2}{(N)(C-1)}}$$

- For a test of independence with two categories per independent variable (2 × 2), the effect measure is called *phi* (ϕ). Phi is interpreted in a similar manner to that of *r*. The formula is as follows:

$$\phi = \sqrt{\frac{\chi^2}{N}}$$

- When a test of independence has more than 2 categories per IV, the above formula becomes somewhat more complicated and uses a new statistic. This statistic is called a Cramer's *V*. This formula includes phi, which is then amended for the number of categories. The formula is as follows:

$$V = \sqrt{\frac{\phi^2}{(\text{Variable with the fewest number of categories}) - 1}}$$

Module 33

- *Power* is the ability of a test to correctly detect an effect (to correctly reject the null hypothesis). In other words, power is the complement of a Type 2 error. The power of a study can be calculated by subtracting 1 from the probability of making a Type 2 error. Increasing the amount of power in a study is very desirable.
- Power is primarily affected by five things: (1) Type 1 error, (2) the directionality of the test, (3) the size of the effect, (4) error variance, and (5) sample size. As alpha increases, the amount of power a study has increases. This is because moving the alpha level closer to the mean (increasing alpha) allows for a larger region of rejection. With a larger region of rejection, the chances of correctly rejecting the null hypothesis increase. This is also how the directionality of a test will influence power, as one-tailed tests have a larger region of rejection in a single tail than a two-tailed test. Similarly, as effect size increases, so does power. A larger effect indicates less overlap between the null distribution and the alternative distribution. The lesser the overlap, the higher the chances of correctly rejecting the null hypothesis (an increase in power). The reverse is also true in that a large amount of error variance will reduce

the amount of power in a study. Finally, large samples increase the chances of rejecting the null hypothesis, and so they are related to an increase in power. Recall that anything that decreases the size of the denominator in a hypothesis test will increase the chances of finding a significant result. Larger samples are related to a decrease in amount of expected error, and thus to an increase in power.

- It is important to note that behavioral research is a complex process, and as such we cannot always expect findings that are as straightforward as an IV being the sole cause of change in a DV. In considering this, it is important to note that as you increase power, the chances of making a Type 1 error also increase.

Learning Objectives

Module 32

- Distinguish between statistical significance and practical or clinical importance
- Understand the influence of sample size on statistical significance
- Know which measure of effect size is appropriate for each test of statistical significance
- Calculate various measures of effect size
- Know the guidelines for interpreting effect size

Module 33

- Know the mathematical relationship between Type 2 error and power
- Understand the effect that various factors have on power
- Understand the interrelationship of statistical significance, effect size, and power
- Use the graphs or tables to determine power, given known values of other variables
- Use graphs or tables to determine necessary sample size for a desired power, given known values of other variables

Computational Exercises

1. A study was comparing a new drug therapy for pain management by comparing those on the new medication with those on no medication. The mean pain rating for those on the new drug was a rating of $M = 67$. The mean pain rating for those not on the drug was $M = 78$. The results for this test were significant at the .05 level, but the researcher is uncertain if this means the treatment was effective. If the overall standard deviation was 7, what is the effect size for this study, and how would you classify it?

2. A pharmaceutical company is interested in developing a new drug for depression. They compare the depression symptoms of those who are not on the medication with those on the depression medication. They obtain a difference in means of 3.1 with an overall standard deviation of 3.7. What was the effect size for this study, and how would you classify it?

3. A scientist is interested in looking at how the brain reacts to stress. He compares the activity of the prefrontal cortex in a sample of 40 individuals who are under stress with that of 40 individuals who are not under stress. For the stress group, he obtains a mean of 25 points of activity. For the nonstress group, he obtains a mean

of 17 points of activity. Although this difference is significant, he is hesitant to report his finding until he has obtained an effect size. If the overall standard deviation was 15.1, how big an effect does being stressed have on brain activity?

4. Researchers are interested in the effects of a new therapy for generalized anxiety disorder (GAD). Unfortunately, they are only able to recruit a handful of participants, $N = 10$. They compare the anxiety of those who received the new therapy ($M = 12$) with the anxiety of those who did not ($M = 18$) and obtain a nonsignificant result at the .05 level. The overall standard deviation for all participants was 5.2. Should the researchers be discouraged with their results?

5. Despite having a good effect size, the researchers in Question 4 are receiving some criticism for their study from the larger psychological community. If they obtained a t value of 6.5, what other measure of effect size could they use to bolster their argument?

6. A new toaster oven is advertising that it can toast faster than any other toaster! You are skeptical when you receive this toaster as a gift and decide to compare it with your already working toaster by toasting $N = 20$ (10 each) slices of bread. The new toaster takes an average of $M = 68$ s to toast bread, whereas your old toaster takes an average of $M = 73$ s to toast bread. You find the standard error to be 2.1. Is this a significant result at a .05 level error rate? What is the effect size for this result? Does the new toaster's faster toasting seem to matter?

7. A researcher is interested in reducing the intensity of seizures in those with seizure disorder. The drug is administered to 15 people with seizure disorder, and another 15 people with seizure disorder are monitored as a comparison. As rated by a third party, the intensity of seizures of those on the medication has a mean of 17. The intensity of seizures of those not on the medication has a mean of 23. If the standard error is 8, what is the significance and effect size of the medication?

8. A researcher is interested in assessing the effect of an herbal supplement on calcium in bones. He decides to create three groups: (1) herbal supplement, (2) placebo, and (3) control group. He obtains the following results:

	SS	df	MS	F
Between	687	2.00	343.50	13.14
Within	784	30.00	26.13	
Total	1471.00	32.00		

What is the size of the effect of the herbal supplement?

9. You are interested in studying the effects of a medication designed to suppress the symptoms of HIV. You decide to compare three medications with a placebo group in order to determine the effect. You obtain the following results:

	SS	df	MS	F
Between	347.00	3.00	115.67	5.36
Within	647.00	30.00	21.57	
Total	994.00	33.00		

What is the size of the effect of the medication?

10. A marketing company is interested in determining the influence of a new advertisement for running shoes. They compare the interest (1 to 10) of those who have seen the advertisement with the interest of those who have not. They also believe that gender will affect the decision such that men will be more influenced by the ad than women. Here are their results:

	SS	df	MS	F
Advertisement	68	1.00	68.00	4.65
Gender	75	1.00	75.00	5.13
Interaction	120	1.00	120.00	8.21
Within	658	45.00	14.62	
Total	921.00	48.00		

What is the size of the effect of gender and the advertisement on interest in the shoe?

11. A new study is being conducted on the influence of driving under the influence of alcohol. The study is comparing the number of accidents had in a driving simulation by those with a blood alcohol level (BAC) of 2.0, 0.5, and none. Here are the results of the study:

	SS	df	MS	F
Between	423	2.00	211.50	5.37
Within	473	12.00	39.42	
Total	896.00	14.00		

What is the size of the effect of alcohol on driving accidents?

12. A store is trying to increase the amount of customer traffic. To increase traffic, the store runs a special sales promotion in which the first 1,000 customers will receive a gift. The store then compares the number of people who entered their store (which was 1,240) as compared with their three top competitors and obtains a $\chi^2 = 7.89$. What is the effect size of this promotion? Did this promotion seem to affect the number of people entering the store?

13. The following week, one of the competitors (Store 2) of the store in Question 12 (Store 1) received only 700 customers during the promotion weekend. To increase customers, Store 2 runs its own promotion the following weekend and obtains 1,231 customers. On this weekend, Store 1 received only 800 customers. An analyst decides to determine if the promotion and store interacted to have a significant increase in customers and obtained a $\chi^2 = 12.54$. What was the effect size for this 2 (store) × 2 (promotion) chi-square? Use information from Question 12 to calculate N. How does this compare with the effect size found in Question 12?

14. In the computational exercises for Part XIII, Question 12 asked whether there was a relation between gender and preference of alcoholic beverage. The $\chi^2 = 40.00$, and there were a total of 100 participants. What was the effect size of this analysis?

Answers to Odd-Numbered Computational Exercises

1. Cohen's $d = 1.57$; large effect.

3. Cohen's $d = 0.53$; medium effect.

5. $r = 0.92$. This is also a large effect.

7. $t(28) = 0.75$, $p < .05$. $r = .14$, small effect. It appears that the medication is not effective.

9. $\eta = .35$

11. $\eta = .47$

13. $\phi = .06$; the effect size is comparable.

True/False Questions

1. A t test will tell you whether or not a new therapy is effective by showing that the therapy group is different from a placebo group.

2. Measures of effect size provide concrete measurements of the size of the effect.

3. A Cohen's d differs from a t test in that it uses a standard deviation in the denominator instead of a standard error.

4. A Cohen's d of 0.95 is a medium effect.

5. An $r = -1$ indicates no effect.

6. The calculation of effect size for a chi-square test depends on which test you have used.

7. Cohen's d and r should provide you with the same exact value for effect size.

8. A Cohen's d of 0.43 and an η of .43 have the same meaning.

9. Power is $1 - \alpha$.

10. Decreasing the sample size will reduce a study's power.

11. A large treatment effect will increase power.

12. Increases in power are always preferred, with a 100% power being ideal.

13. A large enough sample will make even the smallest difference between two means appear significant.

14. The directionality of a test is related to power in that two-tailed tests increase power.

15. Changing the numerator of a hypothesis such that it has a smaller number will decrease power.

Answers to Odd-Numbered True/False Questions

1. False

3. True

5. False

7. False

9. False

11. True

13. True

15. True

Short-Answer Questions

1. What is the next question that is asked after obtaining a significant result? Why is this question important?

2. How can a researcher "force" a result to be statistically significant?

3. How does Cohen's d measure effect size?

4. What about the Cohen's d differentiates it from a t test? How does this change the interpretation and utility of Cohen's d?

5. How does r measure effect size?

6. What is η? Why does the formula include a square root?

7. What does it mean when a test is statistically significant but has a very small effect size?

8. What are ϕ and Cramer's V? When is it appropriate to use either?

9. What is the scale that η, ϕ, Cramer's V, and r use to measure effect size? What are its limits and center point?

10. What is power in a study? Why is it desirable?

11. What are the factors that affect power?

12. Explain how a large effect size and a small error variance work in conjunction to increase power.

13. Explain how changing the α level and the directionality of a test have the same effect on power.

14. Explain the relationship between sample size and power. How does changing sample size change power? How does the size of an effect affect the size of the sample needed to maintain a certain level of power?

15. What is the pitfall of increasing power?

Answers to Odd-Numbered Short-Answer Questions

1. The next question that should be asked is, "Was this difference meaningful? How large an effect did the IV have on the DV?" This question is important because it provides more insight into our research question. Any statistic can be significant with a large enough sample. Effect size helps us determine if the significant difference was meaningful or superfluous.

3. Cohen's d rescales the difference between two means into standard deviation units. This enables you to determine the distance between the two hypothesized distributions in terms of standard deviations.

5. Calculating r gives a measure of effect size that assesses the extent to which two variables, the IV and DV, are related.

7. It indicates that although the chance of obtaining this mean difference by chance is low, the IV had minimal impact on the DV. As a result, the utility of the results of the study may be minimized.

9. The scale that is used by these measures is –1 to 1. The center point is 0, which indicates no relation between the IV and DV.

11. There are five factors that affect power: (1) sample size, (2) error variance, (3) directionality of the test, (4) effect size, and (5) Type 1 error level.

13. Changing the directionality of a test changes the α level in one tail. Thus, these principles are essentially the same as using a one-tailed test, which will increase the region of rejection in one specific tail, increasing α and power.

15. The main issue with increasing power is that it can lead you to make a Type 1 error. Also, it is difficult to state that the result obtained from your study will hold true in all situations. As a result, the extent to which you can correctly reject a null hypothesis in one situation may be very different from another situation.

Multiple-Choice Questions

1. Which of the following does not affect power?
   a. Effect size
   b. α
   c. Sample size
   d. The number of independent variables

2. The chance of making a Type 2 error is .31. What is the power?
   a. .31
   b. .69
   c. .05
   d. .80

3. Which element of power does the researcher have the most control over?
   a. Effect size
   b. α
   c. Sample size
   d. Error variance

4. What is considered a medium effect?
   a. 0.3 to 0.8
   b. 0.4 and greater
   c. 0.2
   d. It depends on the measure of effect size

5. The difference between two means is 7, and the standard deviation for all the scores in the study is 4. What is Cohen's d?
   a. 3.5
   b. 1.75
   c. 0.5
   d. 7

6. An independent-samples *t* test yields a *t* of 4.3 with 16 *df*. What should the researcher conclude about this result (use α of .01)?
   a. The result is statistically significant with a large effect
   b. The result is not statistically significant with a large effect
   c. The result is statistically significant with a small effect
   d. The result is not statistically significant with a small effect

7. What is the effect size of the previous question?
   a. .54
   b. .64
   c. .73
   d. 1.23

8. What can an effect size *r* never be?
   a. −.5
   b. 0
   c. 1.32
   d. .1

9. The results of an ANOVA provide an $SS_{bet} = 582$ and an $SS_{tot} = 1,839$. What is the size of this effect?
   a. .32
   b. .47
   c. .89
   d. 1.2

10. A 3 × 2 test of independence had an $N = 150$ and a $\chi^2 = 5.3$. What is the size of this effect?
    a. .19
    b. .04
    c. .63
    d. .47

11. With the following information, $M_1 = 8.47$ and $M_2 = 10.47$, and an overall $s = 2.68$, what is the effect size?
    a. 2
    b. .74
    c. 1.23
    d. .98

12. As sample size _____, power _____.
    a. Increases, decreases
    b. Decreases, increases
    c. Increases, increases
    d. Remains the same, decreases

13. As Type 1 error rate _____, power _____, and the chances of making a Type 2 error _____.
    a. Increases, increases, increase
    b. Decreases, decreases, decrease
    c. Increases, decreases, increase
    d. Increases, increases, decrease

14. As effect size _____, the sample size needed to obtain adequate power to detect the effect _____.

a. Increases, increases
b. Decreases, increases
c. Remains the same, increases
d. Increases, remains the same

15. In a study comparing the time it takes for two types of paint to dry on $N = 31$ walls, a t statistic of 1.46 is obtained. If the standard error for the two groups is 10.4, which of the following is true?
    a. t is significant at the .05 level and the effect size is $r = .26$
    b. t is not significant at the .05 level and the effect size is $r = .26$
    c. t is significant at the .01 level and the effect size is $r = .07$
    d. t is not significant at the .01 level and the effect size is $r = .07$

16. An electric company is trying to make a brighter lightbulb. Their standard lightbulbs have an average brightness rating of 5.2, and their new lightbulbs have an average brightness rating of 6.9. The standard error between these two groups of lightbulbs is 2.1. If there are a total of 46 lightbulbs in this study, which of the following is true?
    a. t is significant at the .05 level and the effect size is $r = .12$
    b. t is not significant at the .05 level and the effect size is $r = .12$
    c. t is significant at the .01 level and the effect size is $r = .2$
    d. t is not significant at the .01 level and the effect size is $r = .2$

17. Using the following ANOVA source table, what is the effect size for Factor 1 and the effect size for Factor 2?

	SS	df	MS	F
Factor 1	657.00	4.00	164.25	3.32
Factor 2	327.00	3.00	109.00	2.20
Interaction	412.00	12.00	34.33	0.69
Within	3214.00	65.00	49.45	
Total	4610.00	84.00		

a. Factor 1 η = .14, Factor 2 η = .08
b. Factor 1 η = .54, Factor 2 η = .08
c. Factor 1 η = .14, Factor 2 η = .32
d. Factor 1 η = .54, Factor 2 η = .32

18. Using the source table from Question 17, how would you classify each of the effect sizes?
    a. They are all small
    b. They are all medium
    c. They are all large
    d. Factor 1 is large, but Factor 2 and the interaction are small

19. How do small error variances and large sample sizes affect power?
    a. They affect the denominator of a hypothesis test, which lowers power
    b. They affect the denominator of a hypothesis test, which increases power
    c. They affect the numerator of a hypothesis test, which lowers power
    d. They affect the numerator of a hypothesis test, which increases power

20. A researcher is testing a new treatment for post-traumatic stress syndrome, but discovers that his samples have a tremendous amount of variability. What should the researcher worry about doing in this situation, and how could he rectify the problem?
    a. The researcher should be concerned about small effect size. He could rectify this by removing people from the study.
    b. The researcher should be concerned about low power. He could rectify this by removing people from the study.
    c. The researcher should be concerned about a small effect size. He could rectify this by increasing power.
    d. The researcher should be concerned about low power. He could rectify this by including more people in the study.

Answers to Odd-Numbered Multiple-Choice Questions

1. d
3. c
5. b
7. c
9. a
11. b
13. d
15. b
17. a
19. b

Part XV Study Guide

Correlation

Part XV Summary

Module 34

- Correlational studies differ from the work that has been covered thus far. Correlational studies contain a single group and seek only to find relations between variables rather than cause and effect. An example of a relation between variables is a *reliability test*. A teacher could give his or her students two separate tests on the same material and then see if scores on both forms were related. If the scores were, this would indicate that both tests were equivalent. However, if the scores were unrelated, then it may be the case that the two tests were not equivalent. Another use of a correctional study is *prediction*, determining how one will do on Variable A based on one's performance on Variable B. Using the previous example, imagine the teacher wants to determine how his or her students will do on a test later in the course. The teacher could use the scores from the first test to predict scores on the later test, because the teacher can assume that students' test performance stays relatively stable in the same class.
- One method of displaying correlational studies is through a scatterplot. Scatterplots are graphs that display an individual's scores on two separate variables. One variable is represented on the X-axis and the other is represented on the Y-axis. One's scores are graphed by using them as coordinates on the graph. The relation between the two variables is called a *bivariate* relation.
- Correlational studies indicate the strength of a relationship between two variables. The strength, magnitude, or relation is measured on a scale that ranges from −1 to 1, and 1 indicates the strongest possible positive relation. A *positive* relation means that increases on Variable A are related to increases on Variable B. The scatterplot for a positive relationship resembles a straight line starting at the origin of the graph and radiating out at a 45° angle. The closer the relation is to 1, the more the plot will resemble a straight line. Conversely, −1 indicates the strongest possible negative relation. A *negative* relation means that increases on Variable A are related to decreases on Variable B. The scatterplot for a negative relationship resembles a straight line starting at the top of the Y-axis and moving down at a 45° angle until it touches the X-axis. Again, the closer the relation is to −1, the more the plot will resemble a straight line. Finally, a score of 0 indicates no relation. This means that any change on Variable A is unrelated to change on Variable B. The scatterplot for a correlation of 0 would appear very much like a blob. It is important to note that in the social sciences, correlations of 1 and −1 are rarely achieved.

- Correlations explain relations between two linearly related variables. *Linearity* means that the amount of increase in Variable A for a unit increase in Variable B is consistent throughout. When linearly related variables are drawn on a scatterplot, the points resemble a straight line. Variables may also be related in a *curvilinear* fashion. This indicates that an increase on Variable A is related to varying change in Variable B throughout. When curvilinear variables are drawn on a scatterplot, the points resemble a curved line.
- Correlations may be heavily influenced by outliers. *Outliers* are considered extreme scores, or scores that are very different from the rest of the data. An outlier can significantly alter the correlation between two variables.

Module 35

- The strength and direction of a correlation is measured by a *correlation coefficient*. As mentioned previously, strength of a relation is measured on a –1 to 1 scale. There are multiple types of correlation coefficients, but all follow a similar function of measuring the relation between two variables. The correlation coefficient that will be focused on in this book is the *Pearson r*. This is used when the relation between two variables is linear and on either an interval or ratio scale. The raw score formula for a Pearson r is

$$r_{XY} = \frac{N\sum XY - \left(\sum X\right)\left(\sum Y\right)}{\sqrt{\left[N\left(\sum X^2\right) - \left(\sum X\right)^2\right]\left[N\left(\sum Y^2\right) - \left(\sum Y\right)^2\right]}}$$

- There are multiple formulas for a Pearson *r*, one that includes *z* scores, another that includes deviation scores, and finally the one shown above, which includes raw scores. Using the *z*-score formula allows us to graph the *z* scores for each score across a four-quadrant graph. This can be helpful in understanding the direction for positive and negative relations in scatterplots. A positive relation indicates that nearly all the plotted *z* scores will fall in the first or third quadrant, indicating that most of the scores have X, Y coordinates that are either positive or negative. Thus, as scores go up on one variable, they also increase on the other. A negative relation indicates that nearly all the plotted *z* scores will fall in the second or fourth quadrant, indicating that most of the scores have X, Y coordinates that are opposite in signs.
- After obtaining a correlation coefficient, the next question that should be asked is, "Are these variables related or unrelated?" This question is very similar to the question we asked for inferential statistics, "Is this difference between means large or small?" Therefore, correlation coefficients can be tested in a similar process to the one we have used thus far. In this hypothesis test, we are assessing the probability that we would find a relation among these scores. We can expect that if they are not related, the correlation coefficient would be close to 0. We can expect to find some small relation between the variables due to chance, but overall, the scores on the two variables are unrelated. Alternatively, if they are related, we would expect the relation between the variables to be very different from 0.
- Similar to the method used for other inferential statistics, the way we determine a significant relationship is by using a table that can be found in Appendix G. The table is used by first finding the degrees of freedom. The *df* is $N - 2$ (sample size minus the number of pairs of variables). This is used to find the critical value for the correlation. Remember that if the obtained correlation exceeds the critical value, then the null hypothesis is rejected and we have a significant relation. Looking at the table, you may

also notice that there are multiple sets of numbers per *df*, for different Type 1 error rates and directionality of the test. Directionality in a correlation study indicates that the relationship between two variables is expected to be either positive or negative.
- After obtaining a correlation, it is common to write the results in the following format: *r*(*df*) = correlation coefficient, *p* value.

Module 36

- This module focuses on cautions when dealing with correlation. Correlations are among the most commonly used statistics, and it is important to be aware of how they can be misinterpreted.
- The first area of concern with correlational work involves sample size. The chances of rejecting the null hypothesis are reduced as the sample size shrinks. Therefore, you may find that highly related variables do not appear significant because of a very small sample. Alternatively, you may find variables that have very little actual relationship that appear significantly correlated because of a very large sample.
- It is important to remember that statistical significance does not mean that the result is practical. Unfortunately, there are no stringent guidelines to determine when a correlation is practical or impractical. Remember, large sample sizes will cause any correlation coefficient to be significant. Therefore, it is up to you, the user, to determine how much relationship is necessary between two variables to consider the variables related and useful. This will vary greatly depending on the situation. For example, as a teacher, you may want a medium or smaller correlation between student grades as this would show that students are improving over time as opposed to consistently earning the same grade (although that may be good if the students are doing well). Alternatively, if you want to determine if two movie critics are providing similar opinions on movies (determining the reliability of their ratings), you may want a very high correlation.
- Another aspect to consider is restriction of range. *Restriction of range* is when the scores in your sample do not represent the full scale of variability for each variable. Imagine that you have given a test that ranges from 0 to 100. You want to see if scores on this test are related to final grades in the class. You select a random sample from your class and find a very small correlation, .03. You then look at the grades in your sample and discover that they range from 70 to 73. This is a restriction of range. Your small correlation is attributed to the lack of variability in test scores rather than a lack of a real relation.
- The next aspect to consider in a correlation is the homogeneity (similarity) and heterogeneity (difference) of those in your sample, that is, your sample's diversity. This may influence your correlation because, within your overall study, you have distinct subgroups. For example, if we assess the relation between age and basketball performance among high school, college, and professional basketball players, it may appear that there is a strong positive relation that performance increases with age. However, this is more attributed to professional basketball players both being older and having a higher performance ability than the other groups. Within the group of professional basketball players, you may find a very different relation between the variables. Alternatively, homogeneity can decrease the correlation. If you were to look at the relation between income and education in college professors, there may appear to be no relation because everyone is earning similar amounts and has been in school for a similar amount of time.
- Another area of concern for correlational studies is the reliability of your measurements, or how well you are able to assess your variables of interest. Variables that are poorly measured should not be used as indicators. In doing correlational research, it

is critical that you use accurate measurements to draw appropriate conclusions about the relationship between variables. A common saying regarding this principle is "garbage in, garbage out." Poor measurements will give you poor conclusions.
- A common misconception is that the correlation coefficient represents the amount of shared variance between two variables. A correlation of .5 does not mean that 50% of the variance in Variable A is attributed to 50% of the variance in Score B. This is because correlations are in linear units whereas variances are in squared units. To determine the amount of shared variance, we must square the correlation coefficient. Thus, a correlation of .5 indicates that the variables share 25% of the variance.
- The final and most common (and most egregious) mistake made using correlational data is that the relation between the two variables implies causation. Correlations do not imply causation. If two variables are related, it does not indicate that changes in one variable *cause* changes in another. For example, there is a strong correlation between years of education and income. However, going to school longer does not *cause* a pay increase. Rather, going to school longer will make you more competitive in the job market and able to secure a higher paying job. It is imperative that you avoid making this common error when interpreting correlational work.

Learning Objectives

Module 34

- Distinguish between correlation and experimental studies
- Create a scatterplot
- Estimate the strength and direction of a set of data based on its scatterplot
- Understand the effect that outliers have on the strength and direction of a correlation

Module 35

- Distinguish conditions under which the data are appropriately analyzed via a Pearson r versus another correlation statistic
- Determine degrees of freedom
- Calculate a Pearson r
- Use a table to interpret calculated r
- Report results in APA format

Module 36

- Understand the effect that sample size has on the strength of a correlation coefficient
- Distinguish between statistical significance and practical importance
- Understand that guidelines for practical importance differ depending on the use to which the correlation will be put
- Understand the effect that restriction in range has on the strength of a correlation coefficient
- Understand the effect that heterogeneity and homogeneity have on the strength of a correlation coefficient
- Understand the effect that instrument reliability has on the utility of a correlation coefficient
- Distinguish between correlation and common variance
- Distinguish between correlation and causation

Computational Exercises

1. A college admissions officer is interested in determining if the difficulty of a course is related to how well students do in the course. She hypothesizes that these variables will be positively related, because she thinks that a difficult course inspires a student to work harder. Here are the difficulty ratings on a scale of 0 to 20 and the corresponding grade received in the class on a 0 to 100 scale.

Difficulty	Grades
7	77
9	77
12	81
14	88
17	90

   a. What is the Pearson r?
   b. Interpret this correlation, explaining the strength and direction of the relationship. Is there a relationship between course difficulty and course grade?

2. A dentist is interested in determining if there is a relationship between the length of time his patients spend brushing their teeth and the amount of plaque that has built up on their teeth. He obtains the following data from seven of his patients.

Time Brushing	Plaque
2	8
4	7
5	4
5	6
4	10
4	6
1	2

   a. What is the Pearson r?
   b. Interpret this correlation, explaining the strength and direction of the relationship. Is there a relationship between time brushing and plaque buildup?

3. A construction crew is having a competition to determine who can lift the most weight. They believe that the amount a person can lift is highly related to the amount of muscle mass a person has. Here are the data on how much each of the six-person crew lifted and the proportion of muscle mass of each person from his or her most recent physical examination.

Lift Weight	Muscle Mass
152	0.56
176	0.58
180	0.66
228	0.68
263	0.72
307	0.84

   a. What is the Pearson r?
   b. Interpret this correlation, explaining the strength and direction of the relationship.

4. As the editor of a consumer advocate magazine, you are interested in determining if the cost of a blender is associated with its price. Here are the data:

Cost	Quality
31	4
31	2
48	13
31	5
15	4
19	8
38	10
40	13

a. What is the Pearson r?
b. Interpret this correlation, explaining the strength and direction of the relationship. Should you advise readers to purchase a more expensive blender?

5. You are doing a study on eating disorders. You want to determine if there is a relationship between the number of calories a person with binge eating disorder consumes and the amount of hunger he or she feels during the binge. You measure hunger in a self-report rating (1 to 10). Here are the data:

Hunger Rating	Calories Consumed
2	670
7	1,120
0	945
4	551
6	1,362
2	597
3	1,338

a. What is the Pearson r?
b. Interpret this correlation, explaining the strength and direction of the relationship. Does it appear that hunger and calories consumed are related?

6. A magazine recently reported that being attractive was related to the number of times a person was arrested. Below is the scatterplot of these data. Do you think that this is an accurate report? Why or why not?

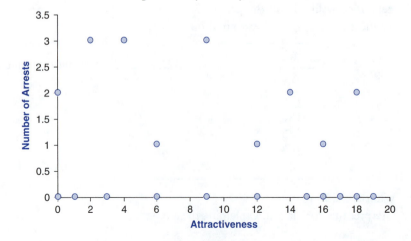

7. A marriage researcher is interested in determining if marital satisfaction among partners is comparable. They obtain the following data from 10 husbands and wives to determine the extent to which they are satisfied with their marriage:

Husband Satisfaction	Wife Satisfaction
2	9
4	34
4	44
7	15
13	19
15	22
25	24
35	11
37	22
43	41

   a. What is the Pearson r?
   b. Interpret this correlation, explaining the strength and direction of the relationship. Does there appear to be a relation between marital satisfaction of the respective partners?

8. Depression and anxiety are considered highly comorbid (co-occurring) disorders in adults. A researcher is interested in determining if this same comorbidity occurs in children. Below are the scores for eight children on a depression and anxiety inventory.

Depression	Anxiety
7	3
8	4
10	6
13	13
13	15
18	24
19	24
19	24

   a. What is the Pearson r?
   b. Interpret this correlation, explaining the strength and direction of the relationship. Does it appear that these symptoms are related in children?

9. Mindfulness is the concept of being present minded in your thinking. Prior research has shown that those with social anxiety have trouble focusing on what is occurring in a social situation, which has been interpreted as a lack of mindfulness. To explore this possible relation, you collect data on both social anxiety and mindfulness. Higher scores indicate more presence of the trait on both measures.

Social Anxiety	Mindfulness
6	8
7	7

Social Anxiety	Mindfulness
8	7
9	5
10	4
11	4
12	3
14	1
14	1

a. What is the Pearson r?

b. Interpret this correlation, explaining the strength and direction of the relationship. Does it appear that these two variables are related?

10. The head of a university is interested in determining how well teacher competency as measured by a standardized test is related to student satisfaction in the classroom. The following data are obtained from five teachers:

Competency	Satisfaction
84	10
57	10
85	1
41	10
43	5

a. What is the Pearson r?

b. Interpret this correlation, explaining the strength and direction of the relationship. Does it appear that there is a relation between teacher competency and student satisfaction?

11. An auto company is interested in determining if the size of a car is related to its gas mileage (in miles per gallon [MPG]). Here are the data they obtain (weight of a car in thousands of pounds).

Weight	MPG
25	38
31	37
31	32
34	32
38	31
40	30
44	27
48	26

a. What is the Pearson r?

b. Interpret this correlation, explaining the strength and direction of the relationship. Does it appear that vehicle weight and gas mileage are related?

12. The department of motor vehicles (DMV) is interested in determining if the amount of time people spend waiting in line is related to the amount of frustration they experience at the DMV. Here are the scores of 8 people waiting in line at the DMV.

Time	Frustration
2	2
9	2
11	4
15	8
23	9
28	9
33	10
65	21

   a. What is the Pearson r?
   b. Interpret this correlation, explaining the strength and direction of the relationship. Does there appear to be a relationship between wait time and frustration?

13. Using the data from the previous question, do there appear to be any outliers in the data? What effect, if any, might these scores have?

14. A graduate training program in clinical psychology is interested in determining if there is a relation between therapist skill and the number of hours the therapist has done therapy. Below are the data from seven student therapists.

Skill	Hours
5	197
7	184
8	218
9	214
12	247
13	232
13	245

   a. What is the Pearson r?
   b. Interpret this correlation, explaining the strength and direction of the relationship. Does there appear to be a relation between therapist skill and hours of experience?

15. A graduate admissions program wants to investigate its recruitment procedure to determine if there is a relation between GRE scores and student performance in graduate school. Here are the GPAs and the GRE scores of 10 graduate students.

GRE	GPA in Grad School
1,160	3.25
1,180	3.14
1,200	3.78
1,280	3.9
1,290	3.8
1,370	4
1,400	4
1,520	4
1,570	3.9
1,590	3.7

a. What is the Pearson *r*?
b. Interpret this correlation, explaining the strength and direction of the relationship. Does it appear that GREs are related to doing well in grad school?

Answers to Odd-Numbered Computational Exercises

1. a. $r = .96$
   b. There is a strong positive relation between course difficulty and grades, $r(3) = .96$, $p < .05$. This indicates that students who do better in courses rate them as being more difficult.

3. a. $r = .96$
   b. There is a strong positive relation between the amount of muscle mass and the amount of weight a person can lift, $r(4) = .96$, $p < .05$.

5. a. $r = .44$
   b. The relation between hunger and calories consumed is not significant, $r(5) = .44$, $p < .05$.

7. a. $r = .07$
   b. There is no relation between marital satisfaction of husbands and wives, $r(8) = .07$, $p < .05$.

9. a. $r = -.99$
   b. There is a very strong negative relation between social anxiety and mindfulness, $r(7) = -.99$, $p < .05$. This indicates that those with more social anxiety have less mindfulness.

11. a. $r = -.94$
    b. There is a strong positive [neg.] relation, $r(6) = -.94$, $p < .01$. As the weight of a car increases, the number of miles per gallon it can go decreases.

13. The last person to provide data appeared to be an outlier and may be inflating the relationship.

15. a. $r = .58$.
    b. The relation between GRE scores and GPA in grad school is not significant, $r(8) = .56$, $p > .05$.

True/False Questions

1. The *df* formula for a Pearson *r* is $N - 1$.

2. Correlational tests can be two-tailed.

3. A statistically significant correlation with a large sample definitely means that the variables are related.

4. A positive correlation indicates that as scores on Variable A increase, scores on Variable B decrease.

5. Two of the uses of a correlation are reliability and prediction.

6. The strength of a correlation refers to the extent that changes in one variable are related to changes in another variable.

7. George notices that he becomes hungry around the same time each day. He concludes that time of day is causing him to be hungry. George is correct in his thinking.

8. A Pearson r measures the strength of two linearly related variables.

9. Outliers can increase the strength of a correlation.

10. A point biserial correlation is the relation between an interval variable and a dichotomous variable.

11. The Pearson r is the only method to assess the relation between two variables.

12. A scatterplot of z scores crosses four quadrants, whereas a scatterplot of raw scores is in only one quadrant.

13. All statistically significant correlations indicate a meaningful relationship among variables and have "real-world" implications.

14. The proportion of shared variance between two variables is r^2.

15. Restriction of range decreases the strength of the correlation.

Answers to Odd-Numbered True/False Questions

1. False
3. False
5. True
7. False
9. False
11. False
13. False
15. True

Short-Answer Questions

1. How does a correlational study differ from a t test or ANOVA?
2. What are two uses of correlational studies? How do they differ in their use?
3. Why are correlations of 1 and −1 rarely found in the social sciences?
4. What does a negative correlation indicate? What does a positive correlation indicate?
5. What is the difference between a linear relation and a curvilinear relation?
6. How does an outlier affect a correlation?
7. What is a Pearson r?
8. What are the hypotheses for correlation coefficients?
9. What is being tested in a hypothesis test for a correlation?
10. How does sample size affect correlations?

11. A teacher asks students at the end of every test to rate the difficulty of the test. She finds that students who rate the test harder tend to do better. Does this mean that a student who rates the test as very difficult will do better? Explain why or why not?

12. What is restriction of range? How does it affect correlations?

13. How do heterogeneity and homogeneity affect correlations? Why is this?

14. In Question 15 of the computation exercises, the relation between GRE scores and grad school GPA was not significant. Looking at the data, what pitfall do you notice that may have influenced this relationship?

15. Why does correlation not imply causation?

Answers to Odd-Numbered Short-Answer Questions

1. A correlation study measures the extent that two variables are related or vary together. An ANOVA and t test measure mean differences.

3. A correlation of 1 or −1 implies a perfect linear relationship, meaning that scores on one variable directly correspond to scores on another. In the social sciences, having such a high level of precision is rare, which reduces the strength of the relationship. Social science research measures tend to contain a fair amount of error.

5. A linear correlation indicates that the relation among the variables follows a straight line, whereas a curvilinear relation indicates that the relation among the variables follows a curve. In a linear relation, a one-unit increase in Variable A is thought to be related to a constant unit increase in Variable B. However, in a curvilinear relation, a one-unit increase in Variable A is related to a different unit increase in Variable B.

7. A Pearson r provides a numerical measure of the strength and direction of a linear relation between two variables.

9. The hypothesis test is determining whether the variation in one variable is related to the variation in another variable. This is done by determining the probability of observing this covariation by chance and then making a statistical decision.

11. No, this does not indicate that giving the test a higher difficulty rating will cause a higher grade in the class. This is an error of inferring causation from a correlation. It may be the case that students who rated the test as very difficult were aware of the complexities of the questions and had to think a great deal to discover the correct answer. Students who did not rated the test as difficult may not have prepared heavily for the test and thus misperceived the test as simple.

13. Heterogeneity of the sample can increase the strength of the correlation. Heterogeneity implies that there are distinct subgroups within the sample that have different levels of ability on the two variables. This can create the false appearance of a strong relation between the two variables. In contrast, homogeneity of the sample can decrease the correlation. This is attributed to the sample being so similar that it fails to use the full scale of measurement.

15. Significant relationships between two variables do not necessarily indicate that one variable causes the other. The precise causal relationship between the two variables is best determined by using additional knowledge about what variables are being correlated.

Multiple-Choice Questions

1. What are the limits of the range of a correlation coefficient?
   a. 0 to 1
   b. 0 to –1
   c. –1 to 1
   d. 10 to 10

2. Bivariate relationships are relationships between
   a. One variable
   b. Two variables
   c. Three variables
   d. Four variables

3. As a correlation increases in strength, the points on the plot begin to _____.
   a. Become closer together, forming a line
   b. Become spread out, forming a "blob"
   c. Fan out at one end
   d. Demonstrate causation

4. A researcher finds that the correlation between stress and intensity of migraine pain is .75. What is the proportion of shared variance among migraine pain and stress?
   a. .75
   b. 1
   c. .25
   d. .57

5. Below are the data representing ratings of attractiveness and the amount of cologne a man was wearing. How would you describe the relationship?

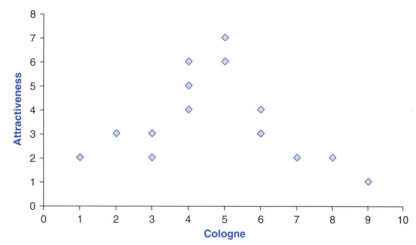

   a. A weak, linear, negative relationship
   b. A strong, linear, positive relationship
   c. A moderate, curvilinear relationship
   d. A weak, linear, positive relationship

6. A researcher obtained a correlation of .6 but found that the result was not significant. The relationship seems to have great practical importance and he believes that the study has led to a Type 2 error. What could the researcher do to obtain statistical significance?

a. Increase the sample size
b. Use different measures
c. Change the hypotheses of the study
d. There is nothing the researcher can do

7. The data below are from a dental study assessing the amount of pain felt and the amount of a new formula for Novocain. What is r?

Pain	Novocain
4	3
1	2
3	1
3	4
2	2

a. .76
b. .23
c. 0
d. .35

8. A heterogeneous group contains subgroups. These subgroups lower the overall correlation between the variables. The existence of the subgroups is an example of what type of variable?
a. Confounding
b. Independent
c. Dependent
d. Causal

9. What type of relationship would you expect between dryness of skin and the amount of moisturizer used?
a. A strong positive relationship
b. A weak positive relationship
c. A strong negative relationship
d. A weak negative relationship

10. One study found that the number of people that drowned per month was significantly correlated to ice cream sales. A friend of yours discovers this finding and tells you not to buy ice cream because it will lead to others drowning. Which pitfall of correlation has your friend succumbed to?
a. Homogeneity of sample
b. Restriction of range
c. Correlation implying causation
d. Common variance

11. How do variables vary in a positive relationship?
a. Increases on one variable are related to decreases on another
b. Decreases on one variable are related to decreases on another
c. Increases on one variable are unrelated to change on another
d. Decreases on one variable are unrelated to change on another

12. Which of the following describes an outlier?
a. Increases the strength of a relationship
b. Decreases the strength of a relationship

c. Can be attributed to a data entry error
d. b and c

13. A Spearman rho measures a relation between two variables on a _____ scale.
    a. Nominal
    b. Ordinal
    c. Interval
    d. Ratio

14. Using a sample size of 15 and a one-tailed test, at what level is a Pearson r of .58 significant?
    a. Significant when $\alpha = .05$, but not $\alpha = .01$
    b. Significant when $\alpha = .01$
    c. Not significant when $\alpha = .05$
    d. Not significant when $\alpha = .025$

15. A researcher is comparing scores on an aptitude test and job performance for 32 individuals. What are the degrees of freedom that should be used when determining if this is statistically significant?
    a. 32
    b. 31
    c. 30
    d. 34

16. Using the information from the previous question, if the researcher obtained a correlation of .43, which of the following statements would be true assuming a two-tailed test?
    a. $r(32) = .43, p > .05$
    b. $r(30) = .43, p > .01$
    c. $r(30) = .43, p < .01$
    d. $r(32) = .43, p < .05$

17. A correlation is reported as $r(17) = .32, p > .05$. How many participants were involved in this study?
    a. 17
    b. 19
    c. 18
    d. 15

18. Two variables share 32% of their variance. What is the Pearson r for these variables?
    a. .32
    b. .57
    c. .10
    d. .64

19. What pitfall of correlation does the "garbage in, garbage out" statement refer to?
    a. Outliers
    b. Unreliable measures
    c. Common variance
    d. Correlation and causation

20. Peggy is a schoolteacher and wants to determine if the age of the students in her kindergarten class is related to their reading ability. Here are the data she obtained. Which pitfall should Peggy be most mindful of in doing this study?

Age	Score
5.3	94
5.6	95
5.4	92
6.1	89
5.9	99

a. Homogeneity of her sample because all her students will be very young
b. Outliers
c. Correlation implying causation
d. Imprecise measures

Answers to Odd-Numbered Multiple-Choice Questions

1. c
3. a
5. c
7. d
9. c
11. b
13. b
15. c
17. b
19. b

Part XVI Study Guide

Linear Prediction

Part XVI Summary

Module 37

- Recall that one of the uses of correlation is prediction, using a person's score on Variable A to estimate what his or her score would be on Variable B. In this situation, we would refer to Variable A as the *predictor* and Variable B as the *criterion*. When the predictor perfectly predicts the criterion, the scatterplot will resemble a straight line and *r* will be 1 (or –1). In this situation, we can determine any value on Variable B as long as we know the score of Variable A. When a score on Variable B (the *Y* variable) is predicted by a score on Variable A (the *X* variable), the score is referred to as *Y'*.
- In the social sciences, variables are rarely perfectly correlated. Thus, when using a score on *X*, there may be many different possibilities for a score on *Y*. This poses a problem as we can only have a single criterion score per predictor score. Probability is used to resolve this situation in that we state that the criterion will predict the most probable single score. All these points form a straight line through the center of all the points of a scatterplot. This line is referred to as a *regression line* or a *best fit line*.
- As mentioned, not all the scores in the scatterplot will fall on the regression line. Sometimes a *Y* value will be above or below *Y'*. The difference between a *Y* value and its *Y'* ($Y - Y'$) is referred to as *prediction error*. This is a measure of how far off the prediction was from the actual value. The function of a prediction score is similar to the function of a mean, and thus the value of the deviations above the prediction score is equal to the value of deviations below the prediction score.
- The prediction error plays a role in selecting where the best fit line will be drawn. The best fit line minimizes the amount of prediction error across all points. The conceptual method that we use to determine how the best fit line will be drawn is as follows: (1) Find the deviation of each raw *Y* from *Y'*. (2) Square the prediction error. (3) Sum the squared prediction error. (4) Do this for all possible lines. (5) Select the line for which the sum of squared prediction error is the smallest. This method is referred to as the *least squares* method because we are selecting the line that has the smallest sum of squared prediction error.
- The mathematical method of calculating a regression line uses the same formula as that for a straight line: $Y = mX + b$, where *m* = the slope and *b* = the intercept on the *Y*-axis. Slope refers to the unit change in *Y* for each unit change in *X*. A slope of 1 indicates that an increase of 1 unit on *X* is related to an increase of 1 unit on *Y*. The formula that is commonly used in statistics is $Y' = bX + a$, where *b* = the slope and *a* = the intercept on the *Y*-axis. Using our knowledge of the relationship between two variables in this equation, the formula changes to

$$Y' = r_{XY}\left(\frac{s_Y}{s_X}\right)X - r_{XY}\left(\frac{s_Y}{s_X}\right)M_X + M_Y$$

- The previous equation, although more intimidating, is equivalent to $Y' = bX + a$. The first term represents the slope and the second and third terms combine to represent the Y-intercept.

Module 38

- This module focuses on calculating confidence intervals for regression lines. A regression line can be considered a number of Y score point estimates for each X score. However, there are a large range of actual Y values for each X, and thus it can be helpful to determine where these scores are expected to fall. Recall that the stronger the correlation, the less the data deviate from the prediction line (i.e., there is a smaller prediction error). In contrast, the weaker the correlation, the more the data will deviate from the prediction line.
- The prediction error for any particular Y variable will be normally distributed. Most scores will deviate a small amount from Y' and very few will deviate a great deal. The extent that a Y score deviates from Y' can be transformed into a standard deviation unit, which is referred to as the *standard error of prediction*. This is the average amount of linear dispersion. The standard error of prediction is calculated with the following formula:

$$s_{YX} = s_Y\sqrt{1 - r_{XY}^2}$$

- Now that we have a standard deviation value for the expected difference from Y, we can calculate a confidence interval. The confidence interval can be calculated using the same proportions as that for the normal curve. In other words, ≈68% of the scores will fall within 1 standard error, ≈95% will fall within 2 standard errors, and so on.
- Only two variables are able to influence the size of the standard error of prediction. This is the size of the standard deviation for the Y variable and the strength of the correlation between X and Y. As the strength of the correlation increases, the standard error decreases. Similarly, as the standard deviation of Y increases, so does the standard error.

Learning Objectives

Module 37

- Understand why a line of best fit minimizes errors of prediction
- Find the equation for the prediction line in a set of bivariate data
- Predict scores on variable Y from known scores on variable X

Module 38

- Know the shape of a distribution of prediction errors
- Calculate a standard error of prediction
- Establish intervals having various probabilities of containing the predicted score
- Understand the effect of sample size on the size of the standard error of prediction
- Understand the effect of variability among predicted scores on the size of the standard error of prediction

Computational Exercises

1. In a regression equation, $b = 0.13$ and $a = 3.30$. Find Y' when X is
   a. 3
   b. 7
   c. 5.43

2. After attending a birthday party for a preschooler and then another for a young adult, you are interested in determining the relationship between age and the number of birthday presents a person receives. After surveying a large number of people and calculating the relationship, you discover that the relationship is .32 with the following data:

	Age	Presents
Mean	15.6	5.4
s	7.3	3.9

   What is the regression line associated with this information? How many presents should you expect on your next birthday?

3. Below are the data obtained from a study that examined the prevalence of PTSD in a population of war refugees. Researchers were interested in determining the extent that exposure to war trauma was related to the severity of the PTSD symptoms. They found a correlation of .23 with the following data:

	Exposure	PTSD Severity
Mean	4.7	11.23
s	2.3	1.9

   What is the regression line associated with this information?

4. A developmental psychologist is interested in determining if the number of hugs children receive per day is related to the amount of affection they feel from their parents. The psychologist obtains a correlation of .65 and the following data:

	Hugs	Affection
Mean	7.8	32.8
s	4.6	7.69

   How much affection should a child that receives 12 hugs a day feel from his or her parents?

5. A parent notices that her adolescent appears to be in a new relationship each week and that her adult child has been in a relationship for many months. She decides to conduct a study to determine if age is related to the length of a relationship in months. She finds a relationship of .35 with the following data:

	Age	Length of Relationship
Mean	25	3.7
s	10.4	0.89

How long should a relationship last for a person who is 15 years old? A person who is 28 years old?

6. Below are the data taken from a nutrition study that attempted to assess the relation between daily calorie consumption and weight. The relationship was found to be .54. Using the additional data provided below, what is the calorie consumption for someone who weighs 189 pounds?

	Weight	Calorie Consumption
Mean	200	1,800
s	30.87	300

7. You notice that the cost of clothing in private boutique stores is much higher than in general department stores. You assume that this is because clothing designers spend more time creating each piece of clothing sold in boutiques than they do those sold in department stores. If the relationship between cost of clothing and the time spent creating it (in days) is .63, what is the regression equation using the following information?

	Cost	Time of Creation
Mean	78	34
s	9.45	4.5

8. What was the 95% confidence interval for the information in Question 2? The 99% confidence interval?

9. For Question 7 (boutique clothing cost), what is the 95% confidence interval for the time of creation for a dress that costs $132?

10. A farmer is interested in determining how many pounds of fertilizer he should use on his potato crop to have them grow to about 3.1 pounds per potato. The relationship between potato size and fertilizer usage is .21. Using the following data taken from numerous farms around the country, how much fertilizer should he use?

	Potato Size	Fertilizer
Mean	4.7	8.7
s	1.2	3.1

11. A neuroscientist is interested in determining how anxiety is related to the amount of GABA in a person's system. He conducts a study and determines that the relationship between GABA and anxiety is −.72. What is the regression line used to predict how anxious a person will be based on the amount of GABA in his or her system?

	GABA	Anxiety
Mean	87.9	25.47
s	20.4	10.23

12. You notice that many of your friends who love music have a large collection of digital music. If the correlation between a person's self-rated love of music and the

amount of digital music he or she owns is .89, what is the formula for the regression line based on the following data?

	Love of Music	Digital Music
Mean	6.5	45.9
s	2.9	7.84

13. How would you classify the relationship between GABA and anxiety in Question 11? What is the 99% confidence interval for each Y'?

Answers to Odd-Numbered Computational Exercises

1.
   a. 3.69
   b. 4.21
   c. 4.01

3. $Y' = 0.19X + 10.34$

5. $Y' = 0.03X + 2.95$.
   15-year-olds should have a 3.4-month relationship. 28-year-olds should have a 3.79-month relationship.

7. $Y' = 0.3X + 10.60$

9. Standard error of prediction = 3.49. For a dress that costs $132, the 95% confidence interval would be $132 \pm (6.99) = 125.01$ to 138.99.

11. $Y' = -0.36X + 57.21$

13. There is a strong negative relationship. The 99% confidence interval is 21.30.

True/False Questions

1. Linear regression lines can be used confidently for curvilinear relationships.

2. A predictor variable causes the criterion variable.

3. When $r = 1$, $Y' = Y$.

4. Stronger correlations reduce the overall amount of prediction error.

5. A best fit line is the one with the smallest sum of squared error for each Y value.

6. Least squares refers to using the regression equation with the least amount of error per X.

7. Comparing typical algebraic notation to statistical notation for regression, $M = a$ and $a = b$.

8. The slope of a regression line is affected by the strength of the relationship between the two variables.

9. The means of each variable are used to calculate the constant of the regression equation.
10. Better measurements of your variables correspond to a more accurate prediction based on a regression line.
11. Regression lines work best for normally distributed data.
12. In the regression equation $Y' = 21.3X + 312$, $Y' = 312$ when $X = 0$.
13. The regression line is a series of point estimates for Y based on each X.
14. Y values are thought to be normally distributed about their corresponding Y' values.
15. The standard deviation of X plays a large role in the standard error of the estimate.

Answers to Odd-Numbered True/False Questions

1. False
3. True
5. True
7. False
9. True
11. True
13. True
15. False

Short-Answer Questions

1. What is a regression line?
2. Why is the regression line like a series of means?
3. What is the least squares method?
4. How does shared variance affect the position of the regression line?
5. How are regression lines different for skewed data?
6. What is the standard error of the estimate?
7. How are confidence intervals calculated for a Y value? Why is this method used or allowed?
8. How does the strength of a correlation affect the standard error? Why is this?
9. How will using a Y score that has a tremendous amount of variability affect the ability of an X score as a predictor?
10. One of the pitfalls of using a correlation concerned using poor measures as they provide poor predictors of other variables. Using your new knowledge of regression lines and prediction error, how would poor measures affect the placement of a regression line?

11. Recall that outliers (extreme scores) can weaken the relationship between two variables. Using a regression line, how would an outlier affect a regression line?

12. How does sample size affect a regression line?

13. A predictor variable is used to estimate a score on a criterion variable. Does this mean a predictor variable causes the score on a criterion variable? Why or why not?

14. Explain why the Y values around each individual Y' score are expected to be normally distributed.

15. What are the two factors that influence the standard error of prediction?

Answers to Odd-Numbered Short-Answer Questions

1. A regression line, or best fit line, is a formula that is used to predict a score on Y using a score on X.

3. The least squares method is the theoretical backbone of a regression line. It states that the regression line is the one that provides the overall smallest sum of squares for the prediction errors.

5. When dealing with skewed data, it may be more appropriate to use multiple regression lines to better predict the criterion variable.

7. Confidence intervals are calculated by multiplying the square root of 1 minus the correlation between X and Y by the standard deviation of Y. This is allowed because it is expected that a Y score will be normally distributed around each Y'.

9. It will reduce the ability for X to predict scores on Y.

11. Outliers would pull the regression line in the direction of the outlier. This will directly affect the slope of the regression line.

13. No it does not. This is an error of correlation implying causation.

15. The standard error of prediction is influenced by the standard deviation of Y and the correlation between X and Y.

Multiple-Choice Questions

1. How many Y' are there per X?
   a. 1
   b. 2
   c. 3
   d. 4

2. How many Y are there per X?
   a. 1
   b. 2
   c. 3
   d. Depends on the data

3. As the strength of the correlation increases, prediction error _____.
   a. Increases
   b. Decreases
   c. Remains the same
   d. Depends on the data set

4. How is the prediction error for any Y score distributed?
   a. Positively skewed
   b. Negatively skewed
   c. Bimodal
   d. Normal

5. For which type of data is a single regression line inappropriate?
   a. Positively skewed
   b. Negatively skewed
   c. Normal
   d. a and b

6. Using the regression equation $Y' = 3X + 2$, what is the prediction error associated with a score of $Y = 12$ and $X = 2$?
   a. 10
   b. 2
   c. 4
   d. 1

7. Using the regression equation $Y' = 2.4X + 12.1$, what is the prediction error associated with a $Y = 5$ and $X = 4$?
   a. 5
   b. 21.7
   c. 3.68
   d. 17.45

8. What is the slope of a best fit line with an $r = .75$, $s_X = 2.1$, and $s_Y = 4.9$?
   a. .98
   b. .32
   c. .47
   d. .51

9. SAT scores and college GPA have a correlation of .25. GRE scores and graduate school GPA have a correlation of .13. Without any additional information, which regression line can you expect to be more accurate?
   a. SAT and college GPA
   b. GRE and graduate school GPA
   c. Both are equal
   d. It depends on the Y-intercept

10. As the standard deviation of Y _____, the accuracy for predicting Y from Y' _____.
    a. Increases, increases
    b. Decreases, decreases
    c. Increases, decreases
    d. Increases, remains the same

Part XVI Study Guide

11. The amount of time spent talking on the phone and the amount of the phone bill have a correlation of .73. Using the following additional information, what is the estimated bill of a person who talks for 320 min?

	Minutes	Bill
Mean	450	98.65
s	57	6.35

    a. 62.05
    b. 25.6
    c. 87.65
    d. 149.70

12. Using the information from the previous question, what is the estimated bill of a person who talks 213 min?
    a. 65
    b. 87.65
    c. 79.09
    d. 62.05

13. Using the information from Question 11, what is the standard error of prediction?
    a. 65.78
    b. 8.65
    c. 1.36
    d. 4.34

14. If two variables are completely uncorrelated ($r = 0$) and use a scale of 0 to 10, what is the standard error expected to be?
    a. 0
    b. 1
    c. 10
    d. Whatever the value of s_Y is.

15. Using the answer from Question 13, what would the 95% confidence interval be for a person that spoke 156 min?
    a. 60–80
    b. 36.48–78.36
    c. 66.3–83.66
    d. 87.36–95.4

16. One of the factors that is related to the size of the standard error is
    a. The amount of relationship between X and Y
    b. The mean of X
    c. The mean of Y
    d. The standard deviation of X

17. A men's suit salesman guesses his customers' waist size based on their height. The correlation between height and waist size is .78. He bases his assumptions on the following data taken from the store's sales history.

	Height	Waist Size
Mean	68	30
s	8	2.8

What is the regression equation that the salesman uses?

a. $Y' = .27X + 11.44$
b. $Y' = 1.27X + 12.44$
c. $Y' = .78X + 15.98$
d. $Y' = 1.3X + 98.7$

18. What is the standard error of prediction for the men's waistline regression equation in Question 17?
    a. 9
    b. 10.23
    c. 1.75
    d. 3.75

19. The salesman from Question 17 grossly overestimates a customer's waist size. The customer was 78 in. tall with a waist size of 28. How far off was the salesman, if he obtained his estimate from the regression equation?
    a. −2.5
    b. −4.5
    c. −9.5
    d. 11.2

20. What would the 99% confidence interval be for a person with a 28 in. waist?
    a. 36 to 29
    b. 27.24 to 37.76
    c. 30.75 to 34.25
    d. 28 to 36

Answers to Odd-Numbered Multiple-Choice Questions

1. a
3. b
5. d
7. b
9. a
11. c
13. d
15. c
17. a
19. b

Part XVII Study Guide

Review

Part XVII Summary

Module 39

- You have now learned many different statistical procedures that enable you to answer a number of questions. Unfortunately, real research questions are not presented as neatly they have been in this book. The biggest challenge when preparing to do an analysis can be determining which analysis will best answer your research questions.
- In determining how to go about selecting the best test, it can be helpful to ask yourself questions about the characteristics of your research question. These include questions about the study's purpose (displaying or reporting information, establishing a relationship, establishing a cause), the scales of measurement that are used, the number of groups you have, and the amount of variables used. The two flowcharts of this module provide good models for the different questions you should ask regarding your analyses.
- Recall that descriptive statistics are used to describe a group of numbers. One of the simplest ways to describe a group of numbers is by creating a table or graph that displays the entire data set. A frequency table shows the number of cases at a given score and can be used to show percentages in relation to the total sample. Graphs provide a visual depiction of the data and enable us to determine its general shape. For single variables, it can be helpful to create a histogram, which displays the range of scores on the X-axis and the frequency of the scores on the Y-axis. For two variables, a scatterplot is used. Individual scores are represented by individual dots that correspond to their score on X and Y.
- Other descriptive statistics can summarize entire sets of data in a few numbers. Some of the most commonly used descriptive statistics are measures of central tendency—the mean, median, and mode. A mode is the most frequently occurring score, the median is the score that occurs at the exact center of the distribution, and the mean is the average score in the distribution. In addition to being the average number, the mean is also the balance point of the data. Remember that in a normal distribution, all the measures of central tendency are equal.
- Dispersion summarizes the extent to which scores vary in a data set. These measures include the range (the difference between the highest and lowest score), the variance (the average area distance from the mean), and the standard deviation (the average linear distance from the mean). Normal distributions contain approximately six standard deviations.

- Raw scores can be converted into standard deviation units. These scores are referred to as standard score. The z score tells us how many standard deviations a raw score is from the mean. Also, the sign of the standard score (+ or −) specifies where the score is in relation to the mean (above or below).
- A correlation provides information about how two variables are related. From a correlation coefficient, we can determine if the variables are positively related or negatively related, and the strength of this relationship. It is important to note that no matter how strongly two variables are related, it does not indicate that their relationship is causal.
- The other types of statistics that we focused on were inferential statistics. Inferential statistics are used to draw conclusions about a population from sample data. Inferential statistics usually determine whether one treatment group's performance differs significantly from that of another treatment group. There are multiple different inferential tests, each appropriate under different circumstances.
- Parametric tests compare sample statistics with population parameters. A one-sample test compares a sample mean with a population mean. If we know the population standard deviation, we can use a normal deviate Z test. If the difference between the means is greater than the standardized random error, then we can assume that the difference is not attributed to chance, and the sample is representative of a different population. A two-sample test assesses difference between two sample means in a manner similar to that of a one-sample test. The two groups are differentiated by their treatments. Hence, significant outcome differences indicate significant treatment effects. Finally, a multisample test simultaneously addresses the differences between multiple groups. A one-sample ANOVA is used when there is only one independent variable with multiple groups. A second independent variable can be used to assess interaction effects. This is referred to as a factorial ANOVA. The purpose of an ANOVA is to compare between-group differences (how differently the groups' outcomes are from one another) with within-group differences (how different the individuals within each group are from one another). As always, if the between-group differences are more than expected (greater than the within-group differences), then the outcome is considered to be significant. After obtaining a significant result, a post hoc test is conducted to determine which treatment group outcomes are significantly different.
- Correlation coefficients can be tested using a method similar to that of mean differences. The test of a correlation coefficient tells us the probability of observing the relationship between two variables. If the probability of observing the obtained relationship is low, then we can conclude that the variables are significantly related. The strength and direction of this relationship is provided by the correlation coefficient.
- Nonparametric tests address variables that are not based on population parameters. The questions we are able to ask with nonparametric tests include those about ranking across groups, similarities in shape of data, and response patterns. The first nonparametric test that was covered was the chi-square test for goodness of fit. This test determined how well the frequencies in one group fit (or were similar to) those of another group. The other nonparametric test that was covered was the chi-square test of independence. This test was similar to the interaction term of a factorial ANOVA in that it sought to determine if two variables acted independently of one another. If the two variables are independent, the relative frequencies among the different groups are expected to be approximately equal. If the observed frequencies are substantially different from the expected values, then we conclude that the independent variables are not independent.

Computational Exercises

1. The following data are taken from a class of statistics students who were asked to rate their enjoyment of statistics on a scale of 1 to 10. Describe these data to a colleague using the mean, the standard deviation, and the shape of the distribution.

Score
99
9
49
97
68
94
77
27
32
62
36
8
65
32
62

2. A professor obtains a sample of responses from her current statistics class regarding their opinion of her statistics course. The responses are below. Describe the data to a colleague using the mean, the standard deviation, and the shape of the distribution.

Score
94
32
7
36
40
66
78
11
74
73
76
32
86
56
87

3. Using the information from the previous two questions, what is the appropriate test statistic to determine if the two classes differ significantly in their enjoyment of statistics? Calculate the test statistic with a Type 1 error level of .05.

4. The professor in Question 2 is now interested in determining how her students feel about other classes they are taking. Surprisingly, all the students in her class are also taking a physics course. Using the information the students provided about their satisfaction with the statistics course (data from Question 2) and the information these same students provided about their satisfaction with a physics course (below), what statistic will test if there is a significant difference in satisfaction between students in the two courses? Calculate and interpret the statistic.

Physics
43
54
26
48
29
47
23
25
28
16
27
41
35
15
29

5. A sports psychologist is interested in determining if those that run in the Boston marathon are younger than marathon runners nationwide. This is because the Boston marathon is considered to be one of the more difficult marathons. The psychologist discovers that the mean age of a marathon runner nationwide is $\mu = 34$ with a $\sigma = 7$. The average age of a sample of 36 runners from the Boston marathon was $M = 29$. What statistic will test if those that run in the Boston marathon are significantly younger than marathon runners nationwide? Calculate and interpret the statistic.

6. You are interested in following up the research that was conducted in the previous question by doing a study to determine whether a group of your friends who are participating in a local marathon are significantly different in age from those participating in the race overall. The average age for the runners in your local marathon is $\mu = 31.4$, but you are unable to obtain a measure of variability. The average age of your $n = 16$ friends is $M = 29.54$ with an $s = 7.98$. What statistic will test if your friends are significantly different in age from the runners in this race? Calculate and interpret the statistic.

7. You are interested in determining if males and females differ on their preference for cognitive therapy as a treatment for generalized anxiety disorder. You describe cognitive therapy to 15 males and 15 females with generalized anxiety disorder to gauge their level of preference for the treatment. You obtain the following scores. What test statistic will test if there is a preference difference between males and females? Calculate and interpret the test statistic at a Type 1 error level of .05.

Females	Males
12	8
12	6
9	10
8	5
11	6
12	8
11	7
6	6
12	4
14	5
11	6
6	10
8	5
11	6
13	10

8. After obtaining the result from the previous study (Question 7), you are interested in determining if their ratings of the treatment will change after being enrolled in the therapy. You decide to focus on the males and obtain the following data after they have completed treatment. Using the *male* data from Question 7 as the pretreatment and the data provided below as the posttreatment, what statistic will test if males' preference for cognitive therapy changes after treatment? Calculate and interpret the statistic.

Pre	Post
8	8
6	14
10	14
5	15
6	8
8	12
7	8
6	12
4	13
5	14
6	8
10	15
5	15
6	15
10	10

9. You are interested in comparing the influence of listening to music while studying for an exam. You test three groups of 10 individuals. The first will listen to rock music, the second will listen to hip-hop, and the third will be a control group and not listen to music. The following are the test scores of all the individuals on the same English final. What statistic will test if listening to music influences test grades? Calculate and interpret the statistic, as well as any additional tests you need to discern the differences.

Rock	Hip-Hop	None
78	72	85
50	68	83
78	68	88
83	81	95
68	80	84
51	71	80
67	66	80
83	75	77
73	76	75
81	74	88

10. You are conducting research on the use of virtual reality (VR) in psychotherapy. You are interested in determining if those who have played video games respond differently to VR therapy than those who have not. You test a total of 50 participants, 25 of whom have played video games and 25 of whom have not. For the group that has played video games, you obtain the following data: $M = 36$, $\sigma^2 = 17$. For the group that has not played video games, you obtain the following data: $M = 28$, $\sigma^2 = 15$. What statistic will test if having played video games affects response to VR therapy? Calculate and interpret the statistic.

11. A team of researchers are interested in determining if a new osteoporosis medication is effective in increasing the density of a person's bones. They intend to do so by comparing the effects of the new medication with an established medication and to a control group. The data obtained are provided below. What statistic will test if there is a significant difference in the effects of the medications on osteoporosis? Calculate and interpret the statistic, as well as any additional tests you need to discern the difference.

New Medication	Current Medication	Control
63	50	30
51	60	33
59	65	33
67	67	45
45	61	41
47	59	47
64	63	17
47	56	17

12. The researchers from the previous question also suspect that the age of the participants may influence the effects of the medication. The researchers have divided their sample into distinct groups, those below the age of 35 and those older than 35. What statistic will test if using the new medication is as effective as using the previous medication and if age affects response to the medication. Calculate and interpret the statistic.

	New Medication	Current Medication
Below 35	63	50
Below 35	51	60
Below 35	59	65

	New Medication	Current Medication
Below 35	67	67
Above 35	45	61
Above 35	47	59
Above 35	64	63
Above 35	47	56

13. A study is comparing the effectiveness of a new drug (Drug A) treatment for treating the symptoms of schizophrenia with the previous medication (Drug B) and to a behavioral therapy. Using the following data of the number of symptoms shown by different schizophrenics given one of the three treatments, what statistic will test how the new medication compares with the previous medication and with behavioral therapy at reducing the symptoms of schizophrenia? Calculate and interpret the statistic.

Drug A	Drug B	Behavior Therapy
22	18	37
23	36	30
28	19	29
16	37	31
22	36	33

14. An ice cream company is interested in determining if the temperature is related to the amount of ice cream they sell. Using the following data, what statistic will test if there is a relationship between ice cream and temperature? Calculate and interpret the statistic.

Temperature	Ice Cream Sold (in Gallons)
84	8
98	7
96	12
91	12
88	9
97	10
67	8
87	12
92	12
103	13

15. A teacher commonly brings in candy for her class. She brings in dark chocolate and milk chocolate and expects that an equal number of the 50 students in the class take one of each type. However, she discovers that 35 students took milk chocolate, while the remaining took dark chocolate. What statistic will test if the observed numbers of selectees are significantly different from what was expected? Calculate and interpret the statistic.

Answers to Odd-Numbered Computational Exercises

1. Mean = 54.47; standard deviation = 30.04; the distribution appears bimodal.

3. Independent-samples t test. $t = -0.05$, $p > .05$. Retain the null hypothesis. There is no difference in enjoyment between the classes.

5. Normal deviate test. $z = 4.27$, $p < .05$. Reject the null hypothesis. Those in the Boston marathon are significantly younger than those in other marathons nationwide.

7. Independent-samples t test. $t(28) = 1.19$, $p > .05$. Retain the null hypothesis. There is no significant gender difference in those who prefer cognitive therapy.

9. One-way ANOVA. $F(2, 27) = 5.26$, $p < .05$. Reject the null hypothesis. Tukey HSD = 9.35. The results indicate that those who did not listen to music did better on the exam than those who listened to any type of music.

11. One-way ANOVA. $F(2, 21) = 21.27$, $p < .05$. Reject the null hypothesis. Tukey HSD = 11.3. The new medication is significantly more effective than taking no medication, but is not significantly different from the current medication.

13. One-way ANOVA. The three treatments differed in their effectiveness at reducing the symptoms of schizophrenia, $F(2, 12) = 3.09$, $p > .05$.

15. Chi-square goodness-of-fit test. $\chi^2 = 8$. Reject the null hypothesis. The amount of chocolate that the students took was significantly different from what was expected.

True/False Questions

1. All parametric tests seek to establish causal relationship between two variables.

2. Descriptive methods seek to summarize information about a large data set in only a few numbers.

3. In a normal distribution, all measures of central tendency are equal.

4. The mean is the best way to describe the central tendency of a bimodal distribution.

5. A Pearson r is used to describe the relationship between an interval variable and a nominal variable.

6. The variance provides a measure of dispersion in linear units.

7. A z score tells the location of a score in a distribution in standard deviation units.

8. Inferential statistics enable you to better describe a set of data.

9. A one-sample t test is used when the population standard deviation is unknown.

10. An independent-samples t test is used when comparing two groups and when both population means are unknown.

11. A factorial ANOVA is used when you have two independent variables.

12. A chi-square test is used when comparing frequencies of nominal variables.

13. A Tukey post hoc test is considered more conservative than a Scheffé post hoc test.

14. Power tells you how big an effect you can expect.

15. A researcher is comparing how well shoe polish works on different types of leather shoes. The researcher compares how well the polish works on sneakers, dress shoes, and sandals. If the researcher were to compare the effectiveness of the shoe polish using 15 different shoes within each category, he or she should analyze these data with independent-samples t tests.

Answers to Odd-Numbered True/False Questions

1. False
3. True
5. False
7. True
9. True
11. True
13. False
15. False

Short-Answer Questions

1. What is the difference between an inferential statistic and a descriptive statistic?
2. What is a standardized score? Give two examples of a standardized score.
3. What is the purpose of a post hoc test?
4. How does an ANOVA (F-ratio test) differ from doing multiple t tests? What is the advantage of using an ANOVA over multiple t tests?
5. Dr. Cho is interested in studying the impact of grandchildren on the happiness of grandparents. He decides to compare happiness ratings of people above the age of 65 with 0, 1, 2, and 3 grandchildren. What type of analysis should Dr. Cho use for these data?
6. You are studying coping strategies in different family configurations. You are interested in determining if responses on a measure of coping differ between single- and dual-parent families. How would you go about analyzing these data?
7. After completing the study outlined in Question 6, you hypothesize that the gender of the child in the family may affect the responses of coping. You decide to incorporate a second independent variable, child gender, into your analysis. How would your analysis change from the response in Question 6?
8. You are interested in examining aspects of relationships. Recently, you have collected data on the number of favors one person in a relationship asks and the amount of frustration experienced by the other person in the relationship. If you were interested in determining how these variables were related, what type of analysis would you conduct?

9. Continuing your relationship research, you now want to determine if there is a significant difference in the amount of time spent watching TV between the members in the relationship. You obtain a measure of how many hours each member of the relationship watches TV and want to determine if there is a significant difference. Keep in mind that in doing this research, you want to compare people within each couple rather than to members of other couples. What type of analysis should you conduct?

10. A professor expects a certain number of his students to obtain As, Bs, and Cs. After completing the term, the professor wants to determine if his estimation was appropriate. What type of analysis should he conduct to determine if he had the correct estimate?

11. A marketing company wants to determine which sport to include in their next advertisement. They ask a focus group to rate their interest in the product with a baseball, basketball, or soccer star in the ad. If the goal of this research is to determine which sport receives the highest ratings, what type of analysis should be conducted?

12. A researcher is interested in determining if obsessive thoughts or compulsive behaviors have a bigger impact on the functioning of those diagnosed with obsessive compulsive disorder. To investigate this matter, the researcher obtains a sample of individuals diagnosed with OCD and asks them to rate impairment in their lives attributed to obsessions and compulsions. The researcher hopes to compare the responses. What type of analysis should the researcher use?

13. Design a study for a 3 × 2 factorial ANOVA.

14. Design a study that would be appropriate for a chi-square goodness-of-fit test.

15. Design a study that would be appropriate for a correlation.

Answers to Odd-Numbered Short-Answer Questions

1. Inferential statistics are used to learn about larger populations from samples. Descriptive statistics are used to describe a set of data.

3. A post hoc test is used to determine which groups are significantly different from one another after obtaining a significant F ratio in an ANOVA.

5. An ANOVA would be most appropriate.

7. A factorial ANOVA would be most appropriate.

9. A related measures t test would be most appropriate.

11. A one-way ANOVA would be most appropriate.

13. Answers will vary.

15. Answers will vary.

Multiple-Choice Questions

1. Which of the following is true of the normal distribution?
   a. Mean > median > mode
   b. Mode > median > mean

c. Mean = median = mode
d. Median < mode < mean

2. In a normal distribution, as a score moves _____ the mean, the probability of obtaining such a score by chance _____.
   a. Further from; increases
   b. Further from; remains the same
   c. Closer to; increases
   d. Closer to; remains the same

3. Which of the following is most appropriate to describe frequency data on a variable?
   a. Means
   b. Variances
   c. Standard deviations
   d. Frequency table

4. After doing poorly on a difficult test, Jack is afraid to tell his parents his grade. What would help his parents put his grade in an appropriate context?
   a. Providing his parents with a frequency table of the grades of all the other students in the class
   b. Telling his parents about how he did in another course in which he is excelling
   c. Telling his parents the mean and standard deviation of the class
   d. Telling his parents how much he loves them

5. You have a friend who consistently procrastinates and is doing poorly in school. You want to show your friend that increasing the amount of time studying is related to better grades. Which of the following analyses would be the most appropriate to prove your point?
   a. Correlation
   b. Chi-square
   c. ANOVA
   d. Normal deviate test

6. You are asked to obtain food orders for a wedding that will include 100 people. There are three options for each guest. How would you best organize this information?
   a. Chi-square goodness-of-fit test
   b. Frequency table
   c. Normal deviate test
   d. Histogram

7. A study is attempting to determine if those with a writing disability do better on a test without a writing component than they do on a similar test with a writing component. Which of the following would be appropriate to examine this hypothesis?
   a. A normal deviate Z test: comparing the test scores of a sample of those with a writing disability to the test scores of the entire population of those with a writing disability.
   b. A repeated-measures t test: comparing the scores of those with a writing disability on tests with and without a writing component.
   c. A one-way ANOVA: comparing the scores on a test of those with a writing disability, those without a writing disability, and those with a reading disability.
   d. A chi-square goodness-of-fit test: comparing the number of those with a writing disability who actually pass the test to the number of those with a writing disability who are expected to pass the test.

8. A clothing company wants to determine if the sports team that is on a basketball jersey influences the number of sales of that jersey. The past 1,000 jerseys sold were from 10 different teams. How should the company go about determining if the team influences the number of jerseys sold?
   a. They should do a one-way ANOVA comparing the frequencies of jerseys sold for each team
   b. They should do a factorial ANOVA comparing the frequencies of jerseys sold for each team
   c. They should do a chi-square goodness-of-fit test, comparing the frequencies of jerseys sold with the expected number of jerseys sold
   d. They should do a chi-square test of independence, comparing the frequencies of jerseys sold with the expected number of jerseys sold

9. A teacher wants to determine if grades for her final have changed between last term's class and this term's class. How should she go about determining this?
   a. An independent-samples t test
   b. A related-samples t test
   c. A normal deviate test
   d. A correlation

10. A phone company is interested in determining when people make the most calls. They obtain frequencies of the number of calls made prior to 9 p.m. and the number of calls made after 9 p.m. They obtain information about the location in which the calls were made, either on the East Coast or the West Coast. If they expect that there should be an equal number of calls regardless of location or time, what tests should they use to analyze the data?
    a. A one-way ANOVA
    b. A factorial ANOVA
    c. A correlation
    d. A chi-square test of independence

11. A therapist is interested in determining if the number of therapy sessions a person has is related to the amount of change in his or her depressive symptoms (as measured on an interval scale). What type of analysis should the researcher use to address this question?
    a. A one-way ANOVA
    b. A factorial ANOVA
    c. A correlation
    d. A chi-square test of independence

12. A researcher rejects the null hypothesis. If he has made an error, what type of error did he make?
    a. Type 1
    b. Type 2

13. A researcher is investigating a neurotransmitter that is involved in the development of new white blood cells. To determine if this neurotransmitter increases white blood cell development, she compares the concentration of white blood cells in rats without the neurotransmitter with the concentration of white blood cells in rats with the neurotransmitter. Which of the following statistics is most appropriate to test her hypothesis?
    a. Repeated measures t test
    b. One-way ANOVA

c. Independent-samples *t* test
d. Factorial ANOVA

14. A school superintendent is determining which of the 15 schools in his district are performing poorly based on a national reading test. He obtains all students' scores from each school. How should the superintendent go about determining which schools are doing significantly worse than the others?
    a. A one-way ANOVA
    b. A factorial ANOVA
    c. A correlation
    d. A chi-square test of independence

15. Which measure of central tendency is most appropriate for positively skewed data?
    a. The mean
    b. The median
    c. The mode
    d. They are all equivalent

16. A Pearson *r* is considered
    a. A descriptive statistic only
    b. An inferential statistic only
    c. Both an inferential and a descriptive statistic
    d. Neither an inferential nor a descriptive statistic

17. Which method is optimal to visually represent a continuous variable?
    a. A histogram
    b. A bar chart
    c. A table of means
    d. A scatterplot

18. In comparing the means of samples that contain different people who are matched on some criterion, you should use a ____ to determine group differences.
    a. Related-samples *t* test
    b. Independent-samples *t* test
    c. Factorial ANOVA
    d. Correlation coefficient

19. What criterion is used to determine when to use a normal deviate test or a one-sample *t* test?
    a. The presence of a population mean
    b. The presence of a population standard deviation
    c. The presence of a sample standard deviation
    d. The presence of a sample mean

20. You are interested in evaluating this validity of this statement: "Absence makes the heart grow fonder." You decide to determine if the length of time people are apart is related to the amount of affection they feel toward that person. You find 40 couples who have been apart for varying amounts of time (continuously scaled) and compare that time with the amount of affection (continuously scaled) they report toward their partner. What statistic should you use to analyze these data?
    a. Related-samples *t* test
    b. Independent-samples *t* test
    c. Factorial ANOVA
    d. A correlation coefficient

Answers to Odd-Numbered Multiple-Choice Questions

1. c
3. d
5. a
7. b
9. a
11. c
13. c
15. b
17. a
19. b